"In this book the author invites his readers to share his own profound journey arising out of melding his formative early experience of wildlife in Africa, contemporary evolutionary accounts and creationist biblical literalism. In clear, coherent and well-argued narratives he takes apart the assumptions common to scientism and creationism and draws on the Christian tradition and biblical sources in order to construct an alternative. This is an intelligently argued yet pastorally sensitive exploration of the challenges faced by evolutionary theists and creationists alike, but its implications go much further than this. For Osborn succeeds in achieving something that few authors manage, namely, a self-critical but compassionate and sometimes humorous account of the difficulties for theists in coming to terms with suffering in the animal world. It deserves to be read and appreciated not just in student courses on God and evolution, but more widely from different ecclesial traditions."

Celia Deane-Drummond, professor of theology, University of Notre Dame

"A beautifully written book! Ron Osborn writes not with spite and ire but with wisdom and generosity of spirit. Where literalism once ruled as the only way to honor Scripture, here the deeper dimensions of God's compassion and sabbath rest come to light. This is the first book I've read on the evolution and creation debate that brought tears to my eyes."

Philip Clayton, Claremont School of Theology, author of *Transforming Christian Theology*

"Ronald Osborn draws together a variety of sources and addresses key issues in this rich project. His analysis of literalism and biblical interpretation is sorely needed in many circles today. And his insights on animal suffering should prove helpful as believers wrestle with the central issues of God's grace in a world of both pleasure and pain, holiness and harrowing abuse."

Thomas Jay Oord, Northwest Nazarene University, Nampa, Idaho

"As religious communities struggle to make sense of their faith traditions after Darwin, they rely on thoughtful and sensitive seers to lead them beyond the shallows of literalism to a deeper encounter with new scientific discoveries. Ronald Osborn's sophisticated reflections on literalism and animal suffering will be helpful to Christians of all denominations who are troubled by the wild ways of evolution."

John F. Haught, senior research fellow, Woodstock Theological Center, Georgetown University

Death Before the Fall

Biblical Literalism and the Problem of Animal Suffering

RONALD E. OSBORN

Foreword by John H. Walton

An imprint of InterVarsity Press
Downers Grove, Illinois

InterVarsity Press
P.O. Box 1400, Downers Grove, IL 60515-1426
World Wide Web: www.ivpress.com
Email: email@ivpress.com

InterVarsity Press® is the book-publishing division of InterVarsity Christian Fellowship/USA®, a movement of students and faculty active on campus at hundreds of universities, colleges and schools of nursing in the United States of America, and a member movement of the International Fellowship of Evangelical Students. For information about local and regional activities, write Public Relations Dept., InterVarsity Christian Fellowship/USA, 6400 Schroeder Rd., P.O. Box 7895, Madison, WI 53707-7895, or visit the IVCF website at www.intervarsity.org.

Cover design: David Fassett
Interior design: Beth Hagenberg
Images: Tiger: DEA / G. DAGLI ORTI/Getty Images
Certificate: ©edge69

ISBN 978-0-8308-4046-5 (print)
ISBN 978-0-8308-9537-3 (digital)

Printed in the United States of America ♾

As a member of the Green Press Initiative, InterVarsity Press is committed to protecting the environment and to the responsible use of natural resources. To learn more, visit greenpressinitiative.org.

Library of Congress Cataloging-in-Publication Data

Osborn, Ronald E., 1975-
Death before the fall : biblical literalism and the problem of animal suffering / Ronald E. Osborn.
pages cm
Includes bibliographical references and index.
ISBN 978-0-8308-4046-5 (pbk. : alk. paper)
1. Bible—Evidences, authority, etc. 2. Bible—Criticism, interpretation, etc. 3. Animal welfare—Religious aspects—Christianity. I. Title.
BS480.O845 2014
231.7'652—dc23

2013046584

P 21 20 19 18 17 16 15 14 13 12 11 10 9 8 7 6 5 4 3 2
Y 32 31 30 29 28 27 26 25 24 23 22 21 20 19 18 17 16 15 14

Contents

Foreword by John H. Walton 7

Introduction 11

Part One: On Literalism

1 The Creation 25
A Plain Reading

2 What's Eating Biblical Literalists? 39
Creationism and the Enlightenment Project

3 Unwholesome Complexity 49
Literalism as Scientism's Pale Mimetic Rival

4 Progressive Versus Degenerating Science 59
Weighing Incommensurable Paradigms

5 Does Your God Need Stage Props? 68
On the Theological Necessity of Methodological Atheism

6 The Enclave Mentality 76
Identity Foreclosure and the Fundamentalist Mind

7 The Gnostic Syndrome 86
When Literalism Becomes a Heresy

8 Four Witnesses on the Literal Meaning of Genesis 96
Barth, Calvin, Augustine and Maimonides

9 If Not Foundationalism, What Then? 113
From Tower Building to Net Mending

Part Two: On Animal Suffering

10 Stasis, Deception, Curse 126
Three Literalist Dilemmas

11 A Midrash 140
C. S. Lewis's Cosmic Conflict Theodicy Revisited

12 God of the Whirlwind 150
Animal Ferocity in the Book of Job

13 Creation & Kenosis 157
Evolution and Christ's Self-Emptying Way of the Cross

14 Animal Ethics, Sabbath Rest 166

Conclusion 177

Notes. 181

Subject and Author Index. 193

Foreword

❧

In the current climate of conversation concerning science and faith, many issues are brought under scrutiny. Some involve the rapid advances in genetics and others the application of ancient Near Eastern literature to the interpretation of Scripture. Evolutionary biology, physical and cultural anthropology, neuroscience, quantum mechanics, theology and hermeneutics all have their contributions to make to the complex discussion. Many disputes concern how substantial the scientific conclusions are, while others focus on how we should interpret the Bible and what claims it makes.

One of the most commonly discussed topics is what it means to interpret the biblical text literally. In this book, Osborn does readers the great favor of delving deeply into the issue of what he labels "literalism" to demonstrate how such a hermeneutic can be, and has been, abused by interpreters who have other agendas. Many of those who contend for narrow literalism are quite selective in what they read literally and, in the end, fall prey to the post-Enlightenment demands of scientism. Osborn contends that narrowly literalist scientific creationism not only fails to offer a sound reading of the text but actually damages the text.

A second topic that, in contrast, is little discussed in the modern conversations is animal suffering, despite the fact that many would identify it as the main reason that they cannot accept evolutionary theory. Osborn's case is that while there are no easy answers, those who believe in literalism and creationism are not in any better position on animal suffering than those who accept evolutionary models. In other words, evolutionary theory is not the problem. This is an important point to be made for those who have used animal suffering as the single most important reason that evolution cannot be considered.

This powerfully written book is eminently worth reading for persons of all beliefs or none who want to think deeply about some of the issues in science and faith that perplex us today. Osborn's insights can bring us yet another step along the path toward finding peace from the animosity and angst that surround these issues. But beyond that, he calls us to be better conversation partners; better people. And that is always a message worth listening to.

John H. Walton

Bless the Lord, *O my soul! . . .*

You appoint darkness and it becomes night,

In which all the beasts of the forest prowl about.

The young lions roar after their prey

And seek their food from God.

When the sun rises they withdraw

And lie down in their dens.

Man goes forth to his work

And to his labor until evening.

O Lord, *how many are Your works!*

In wisdom You have made them all;

The earth is full of Your possessions.

There is the sea, great and broad,

In which are swarms without number,

Animals both small and great.

There the ships move along,

And Leviathan, which You have formed to sport in it.

They all wait for You

To give them their food in due season.

You give to them, they gather it up;

You open Your hand, they are satisfied with good.

You hide Your face, they are dismayed;

You take away their spirit, they expire

And return to their dust.

You send forth Your Spirit, they are created;

And You renew the face of the ground.

Psalm 104:1, 20-30 nasb

Introduction

As a child growing up to missionary parents in Zimbabwe not long after its independence from the apartheid regime of Ian Smith's Rhodesia, death in nature was something I had been exposed to from an early age, albeit not in everyday life. My family's home was in a quiet suburb of the serene and modern city of Harare, famous for its well-groomed golf courses, botanical gardens, and wide boulevards lined with jacaranda and flame trees that bloomed spectacular shades of purple, white or orange depending on the season of the year. Here the most deadly animal one might encounter was a mamba, boomslang or other poisonous snake, although sightings of even these were extremely rare. Urbanization and farming had long since driven Africa's famed wildlife far from the places most Zimbabweans lived. Harare was, at least to this child's eyes, a tranquil paradise with endless adventures to be had with my classmates on sunny afternoons as we roamed the city on our bicycles after shedding the gray knee socks and cramped black shoes of our school uniforms—the cruelest legacy, I callowly assumed at the time, of British colonialism. I had much to learn. Often, though, my parents would load my two sisters and me into our mechanically challenged diesel Mercedes sedan, and we would leave the temperate plateau on which Harare sits to camp some 250 miles to the north at a place known as Mana Pools.

Mana Pools is a remote wildlife preserve and UNESCO World Heritage

Site with some of the most spectacular game viewing anywhere in Africa's southern hemisphere. It was one of my family's favorite retreats during the years we lived in Zimbabwe. After passing the final tsetse fly control station on the main road—a low mud hut from which a man in a crumpled khaki uniform would sleepily emerge, armed with a rusty tin canister to spray the undercarriage of our car with a noxious-smelling liquid intended to prevent the deadly insect from returning with us—we were truly into the wild, or into the *bundu,* as it is called in bantu slang. (One of my prized possessions during my elementary school years was a worn copy of the 1967 classic survival manual *Don't Die in the Bundu,* by Col. D. H. Grainger of the Rhodesian Army. I never had any reason to put into practice the lessons in this book, but they provided rich fodder for fantasies of heroic feats of a young boy alone against Africa's elements armed with nothing more than his trusty Swiss Army knife.)

The final leg of the journey into Mana Pools follows a desolate, unpaved track surrounded by thick brush and dotted by colossal baobab trees that look like the remnants of a lost Jurassic Park. One might drive for hours on this path without spotting another vehicle or human being. The first time my family entered this dusty road, however, we had not gone far before we encountered a fresh lion kill. Three young females had taken down a Cape buffalo, which they had not yet dragged into the cover of the bush. Its legs were splayed at odd angles and its side was opened, exposing an impressive rib cage in shades of white and crimson. The lions were feasting on the carcass in the middle of the road, panting heavily as they tore into its body, their chests and muzzles soaked in blood. The air was filled with the stench of death. My father turned off the car engine, and we sat in awed silence watching them feast at a distance of several meters. At last we continued on our way, dipping into the sandy shoulder of the track as we navigated about this scene of beautiful carnage. The lions paid us little notice. I was nine or ten years old at the time, but the memory is still vivid.

It was usually late in the afternoon when we would at last arrive at the Mana Pools campsite, which lies beneath a grove of acacia trees on a high bank overlooking the Zambezi River. After darkness had settled, someone might shine a powerful spotlight or "torch" over the nearby marshy inlets, and we would watch the gleaming eyes of the crocodiles blink and then eerily vanish beneath the water like dying stars. We would lie still in our tents as the campfire dwindled to embers and listen to the haunting cries of jackals in the distance.

There is another distinct sound I remember from those nights as we lay in our beds in the *bundu*. There were no fences or walls in Mana Pools, and wild animals often made their way through camp. Our most common visitors (apart from the kleptomaniac vervet monkeys always seeking a chance to seize unattended food) were herds of elephants. We could often hear their soft tread and breathing just the other side of our tent flaps as they picked up the pods that fell from the acacia trees we slept under. They would frequently pass close enough to step on our tent ropes, although even in the black of night they had an uncanny awareness of the lines and never so much as grazed them. Before dawn, my parents would already be waking my sisters and me for a new safari, since the early hours of the day when the air was still cool and crisp were the best time to spot rare animals. All around us was a world that was deeply mysterious, untamed, dangerous, beautiful and good, waiting to be explored. And the danger was part of its goodness and its beauty.

Herein lies the central riddle of this book. One might, of course, imagine other worlds in other universes without predatory creatures such as crocodiles and lions, and these might be very good and very beautiful worlds as well. But the particular goodness and beauty of Africa's wild places that were such an important part of my childhood were inextricably linked to cycles of birth and death, as well as suffering, ferocity and animal predation. One need not subscribe to seventeenth-century German philosopher Gottfried Leibniz's claim (satirized by Voltaire in *Candide*) that the world as we know it is "the best of all possible worlds" to nevertheless see that adjectives such as *evil* and *cursed,* when applied to the realities of life and death in the animal kingdom, somehow just do not ring true. Mana Pools was *very good*—its lions, jackals, leopards, fish eagles and cobras included. Yet Mana Pools, as a microcosm of nature as a whole, was also an untamed and even unremittingly harsh world, a sealed economy in which all of life was, in the final analysis, sustained by the deaths—often in spectacular and prolonged ways—of other creatures. There is a doubleness to all of animal existence, extending right back to the very beginning as far as we can tell, with birth and death, comedy and tragedy, suffering and grandeur, appearing as the interwoven and inseparable aspects of a single reality that defies easy moral categorization.

For believers in the God of Jewish and Christian Scripture, this poses a grave theological and moral dilemma that is different in kind from the problem of evil arising from the exercise of human free will. It also distinct from (if perhaps related to) the problem of "natural evil" posed by geological upheavals that take human lives, such as the 2010 earthquake in Haiti and the tsunamis in Indonesia and Japan. Simply stated, the trouble is this: Animals, as far as we know, do not have the capacity for anything approaching human moral reasoning and will never be able to comprehend their own suffering in metaphysical or theological terms that might give that suffering meaning *for them*. Why, then, would a just and loving God—not the impersonal spirit of Hegelian idealism that achieves its final ends through the violent dialectics of "history as slaughter-bench,"[1] nor the divinity of Hindu belief who is at once Brahma the creator and Shiva the destroyer of worlds, but the undivided and good Creator God of the Hebrew Bible and New Testament—require or permit such a world to exist? This world is one in which the harrowing suffering of innocent creatures through the violence of other creatures appears at once fraught with terrible savageness and at the same time part of an order that is delicately balanced, achingly beautiful and finely tuned to sustain tremendous diversity of life. If there is a rationally discernible "intelligent design" to the natural world as some believers claim, should we not conclude that the design reveals a pitilessly indifferent if not malevolent intelligence? Why is it that creationists who read "design" from the surface of nature never rhapsodize about the wondrous, irreducible complexity of AIDS viruses, or tapeworms, or serrated shark teeth tiered five rows deep? "It is as if the entire cosmos were somehow predatory," writes Eastern Orthodox theologian David Hart, "a single organism nourishing itself upon the death of everything to which it gives birth, creating and devouring all things with a terrible and impassive majesty."[2]

The pervasiveness of terrible, impassive and majestic forms of predation in nature extends from the great carnivores at the top of the food chain—the prides of lions that have been captured on film killing baby elephants in scenes of such protracted agony that I find them impossible to watch—to the tiniest of insects in the most seemingly tranquil English garden. Aphids would multiply exponentially to unfathomable numbers if not devoured continuously by other creatures. Parasitic wasps hatch thousands at a time to feed with ravenous hunger inside their hosts. Death in nature "is more than extravagance," Annie Dillard concludes, "it is holocaust, parody, glut."[3] Dillard recounts the "unholy revulsion" of devout nineteenth-century French naturalist J. Henri Fabre after he observed a macabre drama unfold between a bee, a Philanthus wasp and a praying mantis. The wasp paralyzed the bee and squeezed its belly so that it would disgorge its sweet syrup, feeding by licking the dying insect's tongue. At this moment a mantis suddenly appeared, pinned the wasp and began to munch its abdomen. As the wasp was itself being eaten, it continued to lick the tongue of the dying bee, unable to forgo the sweetness of the honey even as it writhed in its own death agony. "Nature is, above all, profligate," Dillard writes. "Nature will try anything once. This is what the sign of the insects says. No form is too gruesome, no behavior too grotesque. . . . This is a spendthrift economy; though nothing is lost, all is spent."[4]

Yet in another light, the bee, the wasp and the mantis are not grotesque creatures at all but rather intricately beautiful and noble actors who faithfully fulfill their mysterious parts in the drama of life. The deaths we observe in nature in the present, scientists tell us, are only the tiniest part of a process that has extended over many millions of years; more than ninety percent of all the species that have ever appeared on earth are now extinct.[5] Strictly speaking, though, the problem of nature's doubleness is not one of evolution as such. It is the problem of existence. Why *this* world in *this* form rather than another?

Like millions of Christians, I was raised to believe that God created all of earth's creatures in six literal days in the relatively recent past. In the beginning, there was no mortality and no predation of any kind. The natural world—my parents, pastors and elementary school teachers all sincerely believed and taught me—was radically altered as a result of Adam and Eve's decision to eat the forbidden fruit. The blame for all death and all suffering in nature thus fell squarely upon rebellious humans. This was why lions now killed Cape buffalo in Mana Pools and why there were crocodiles and bilharzia parasites in the Zambezi and Limpopo rivers. But of course no human action could have produced such an instantaneous change, not simply in the instincts but also in the anatomical structures of countless creatures. The idea that the lions in Eden were docile vegetarians with dagger-sharp claws originally designed by God for tearing the bark off trees appeared downright silly. Somehow those massive canine teeth and retractable claws for taking down living prey had *got there*. This seemingly left only one possibility: God himself was responsible for the transformation of all of nature in what amounted to a hostile second creation after Adam and Eve's fall. All mortality and all predation in the animal kingdom were the result of a divine punishment or "curse." The vexing question of the justice of such an act—of why God would inflict death and suffering on innocent creatures to punish sinful humans—did not enter my mind as a child. I simply assumed that older and wiser people whom I loved and trusted had done the hard theological work, and that there were no deeper questions about the creation left to be asked. The task of believers was not to raise difficult problems but to provide confident answers.

Thankfully, my parents never stifled my own religious questioning, and the conversations around our family dinner table were always free and open. Nevertheless, biblical literalism, creationism and the importance of "simple faith" in God's Word were powerfully impressed on my mind in countless ways from my earliest memories as a child. These memories include a vividly illustrated set of children's books—*Uncle Arthur's Bedtime Stories*—written in the 1950s by one Arthur Stanley Maxwell to instill virtues of honesty, obedience and trust in a personal and loving God who intervenes constantly in the lives of those who remember to say their nightly prayers. It was only much later in life, as I began to seriously wrestle with the problem of suffering in Jewish Holocaust literature as an English major at a small Christian liberal arts college in New England, that I truly began to grasp the weaknesses and inadequacies of many of the belief structures in which I had been raised. Or, rather, I came to see how these structures had instilled important beliefs in me at a certain stage in my religious development, but also how the world of Uncle Arthur is a world one can only remain permanently attached to by sealing oneself in a comforting but finally deadening mental and theological cocoon.

The somewhat peculiar religious community in which I was raised and continue to find Christian fellowship, I should perhaps state here, is the Seventh-day Adventist Church, which traces its roots to the Methodist as well as Anabaptist traditions and has long been concerned with the so-called theodicy question—the problem of how a perfectly good and all-powerful God can allow evil or suffering of any kind to exist. Most Adventists are also strongly—at times, unfortunately, stridently—committed to a highly literalistic way of reading the creation story in the first chapters of Genesis. This is not a coincidence. As one Adventist official declared in an angry attack on theistic evolutionists within the Adventist community (an admittedly small minority), individuals who allow for evolutionary ideas in their worldviews

> don't worship the God of the Bible, for that God didn't use a long, protracted, and vicious dog-eat-dog, survival-of-the-fittest paradigm—one that goes against everything He has taught us about love and self-sacrifice—and then lie to us about it by claiming He created life here in six days when He didn't.[6]

As a result of their concern not only for the authority of Scripture but also for the theodicy problem posed by evolutionary biology, Adventists have played an important role in the dubious project of "creation science," beginning with the tireless efforts of self-taught creationist George McCready Price (whose now thoroughly discredited ideas about geology were relied on by William Jennings Bryan in the infamous 1925 Scopes "Monkey Trial," and who helped inspire modern American creationism via the work of John Whitcomb and Henry Morris in the 1960s[7]). This book, which has grown out of a series of occasional articles first published by *Spectrum Magazine* online in 2010 and 2011, is thus to a large extent an open letter—in the form of a somewhat eclectic mustering of arguments, ruminations, interventions and investigations—addressed to the troubled Christian community of which I am myself a part, a community that now finds itself in a state of increasing turmoil among many of its members over questions of faith, theodicy and evolutionary science (with some church officials attempting to resolve the tensions once and for all by turning strict biblicism or literalism on Genesis into a dogmatic litmus test of "true" Adventist identity).

Yet these reflections are not only for or about Adventists (who by the year 2020 may number as many as forty million adherents and regular church attendees worldwide, mostly in Africa and Latin America). Many Christians of all denominational backgrounds are wrestling with the great attrition rate of their young adults as they head off to colleges and universities where they will be exposed to a bewildering array of ideas that are often openly hostile to religious faith. In the face of the challenges of secular modernity, including the corrosive claims of philosophical materialism or scientism, the natural instinct of many devout believers in the US and elsewhere is to try to shore up traditional beliefs through increasingly narrow doctrinal formulations that might restore a lost sense of certainty in an increasingly uncertain world. This instinct has often translated into strict literalism on Genesis and the mental gymnastics of young earth or young life creationism. A 2010 Gallup poll found that fully forty percent of people in the US subscribe to highly literalistic readings of Genesis and creationist ideas.[8] Among evangelical Christians, the percentage is much higher. The move toward "scientific" creationism, however, simply drives many thoughtful young people and others away from religious belief all the faster. When scientifically trained individuals conclude that creationist apologetics are not intellectually credible, Old Testament scholar John Walton notes, "they have too often concluded that the Bible must be rejected."[9]

These tragic decisions against continued participation in religious community from a sense of intolerable cognitive dissonance as a result of what Scripture is assumed to teach—without any room for honest questions, openness to the weight of empirical evidence or new interpretations of the biblical narratives—are not the casualties of evolutionary science. They are, I seek to show from multiple perspectives in part one of this book, "On Literalism," a senseless loss of engaged minds from the body of Christ as a result of problematic philosophical reasoning, the exclusionary logic of fundamentalism and the incoherencies of wooden literalism itself. Christians can and must do better.

Any nonliteralistic approach to Genesis and questions of origins must, however, respond to the obvious moral hazards posed by evolutionary theory. My critique of literalism in part one of this book, although filling the most pages, is to a large extent prolegomena to what I want to say about the theodicy dilemma of animal suffering and mortality in both literalistic and evolutionary paradigms. As Catholic theologian and theistic evolutionist John Haught writes, strict Darwinism's depiction of all of life as nothing more than the final outcome of a purposeless and ruthless struggle for scarce resources "exposes a universe apparently untended by divine compassion."[10] It is to this challenge that I turn in part two, "On Animal Suffering."

While there is a large body of literature providing strong theological grounding for environmental ethics, there are surprisingly few books in print on the topic of animal predation as a distinctive theodicy problem, even though the suffering of animals may be the most severe theodicy dilemma of all given the fundamental questions it raises about God's character as Creator. As I write, Google Books (an imprecise but still helpful index) lists only sixty titles ever written that include somewhere within their pages reference to "animal suffering," "evolution" and "natural evil." Of these works, thirty-seven have been published since 2000. If "theodicy" is substituted for "natural evil," the number of titles shrinks to fifty-two, with thirty-four published since 2000.

My hope, then, is that this book will contribute in a modest way to a small but growing body of Christian reflection on a significant theological riddle. I write not only for those predisposed to agree with me and do not expect—or for that matter desire—to win all readers over to all of my views; I can think of nothing duller than a world in which all critical discussion comes to an end! I hope, though, that whatever conclusions readers take from this book will be the conclusions of careful wrestling with what I have actually written, difficult though this may be given the unavoidably controversial nature of some of what I am constrained to say.

Readers should be fairly warned from the outset that I offer few confident answers to the problem of animal suffering in the manner of some Christian apologists. Indeed, I usually find such "answers" to be morally repellent in the face of the challenges, which I take to be insoluble this side of the parousia (and quite possibly the other side of the parousia as well[11]). Nevertheless, it would be just as morally irresponsible to abandon the search for clues to the theodicy dilemma from a foregone conclusion that the search can yield no answers. The ideas presented in these pages are offered in an open-ended, exploratory form based on the belief that partial answers do exist.

My goal is not to exhaust possibilities but to provoke honest even if unsettling conversations as one member in the body of Christ addressing others. I make no attempt to survey all of the available positions, and I realize that the topics in each of these chapters deserve their own book-length scholarly treatments. I have emphasized the ideas of highly traditional believers in part because I share much of their traditionalism and in part because I want to demonstrate to literalists that one can be a thoroughly orthodox Christian and embrace evolutionary concepts without contradiction.

As a result, I have no doubt failed to do justice to the work of other important thinkers, whose ideas I am by no means closed to. Still, I hope these reflections might provide a helpful even if incomplete map through forbidding terrain, marking some of the pitfalls in literalist accounts—especially when these accounts are turned into religious dogmas—and highlighting other paths that now hold greater theological promise.

The epistemology that informs my "method" (insofar as I have one) follows what has come to be known as the Wesleyan Quadrilateral, so named because it reflects the ideas of Methodist reformer John Wesley. I believe in the paramount authority of Scripture in matters of faith, illuminated though not bound by the interpretive traditions of the church across time, which must be continually tested in the light of both reason and experience to discern present truth. I am enough of a postfoundationalist, though, to see the difficulties in this tidy division of labor. All knowledge comes to us through a complex web of interpenetrating and interwoven channels that are best grasped holistically, as a unified field. By loose analogy, we do not say that a person's mind is the "foundation" on which their physical body and life history are only secondarily built. We say that human personhood involves multiple dimensions of being, none of which can be imagined (although some have tried) apart from the others even if we might at different times place greater emphasis on one or another aspect of a person's being. This is also how we ought to think about the interrelations between revelation, tradition, reason, scientific observation and experience. If there is a reading of Scripture that somehow stands pristinely on its own apart from all of the messiness and contingency as well as all of the richness of human culture and tradition, reason, observation and experience, it is one I have yet to find.

But I am already far ahead of myself. The best place to begin is, of course, at the beginning.

Part One

On Literalism

1

The Creation

A Plain Reading

❧

In the magisterial language of the King James Bible, "In the beginning, God created the heaven and the earth."[i] Genesis 1:1 is, however, an ambiguous text that can be translated as the start of a process rather than a fait accompli. In the Jewish Publication Society translation (following the great medieval rabbi Rashi) it is rendered, "When God *began* to create heaven and earth . . . " Whichever way the verse is read, there is the strong suggestion of a significant time gap between the creation of basic matter and the creation of life and its supporting structures. The earth was "without form and void," the text reads, and "the Spirit of God moved upon the face of the waters."

The theme of creation as process as well as event is continued throughout the narrative in richly suggestive ways. In Genesis 1:3, God says, "Let there be light," and immediately "there was light." The creation of light is thus *ex nihilo,* instantaneous and strictly by divine fiat. The Lord spoke and it was so. However, when we arrive at Genesis 1:11, we find that God does not only create *ex nihilo*. He recruits and involves what he has already created in the next acts of the unfolding drama. "Let the earth bring forth grass, the herb yielding seed, and the fruit tree yielding fruit after his kind, whose seed is

---

[i]All biblical quotations are from the King James Version unless otherwise noted.

in itself," God says. And "the earth brought forth" vegetation. The earth itself therefore participates as an obedient servant to God in the creation process/event. Life is a gift of God. At the same time, the language is clearly focused on natural generation and conveys a strong impression of organic emergence. God commands, but something very different is under way than in the creation of light in verse 3. The earth is charged with a task. The earth brings things forth.

In Genesis 1:22, the text suggests an incomplete or still-empty creation with ecological niches waiting to be filled by living creatures—again, not instantaneously or by divine fiat as in the case of light, but by the animals themselves through procreative processes that will extend across time. The birds and creatures of the seas are commanded by God to "Be fruitful, and multiply," to "fill the waters in the seas" and to "multiply in the earth." The text does not restrict the multiplication of animals to quantitative multiplication alone. We are left entirely free to think that the Creator might be delighted to see his creation multiply not only in number but also in kind. God's way of creating is therefore organic, dynamic, complex and ongoing rather than merely a sequence of staccato punctuation marks by verbal decree. The God of Genesis recruits the creation as a coparticipant in his work as it unfolds, so that not all of the earth is out of nothing or by unmediated speech acts.

In fact, the number of verbs used to describe God's creative activity in the highly compressed narratives of Genesis 1–2 is astounding and by no means limited to the Hebrew word "create." The New American Standard Bible renders these verbs as follows: God "creates" (Gen 1:1, 21, 27). He "moves" (Gen 1:2). He "says" (Gen 1:3, 6, 9, 14, 24). He "separates" (Gen 1:4, 7). He "calls" (Gen 1:5, 8). He "makes" (Gen 1:7, 16, 25; 2:4). He "gathers" (Gen 1:9). He "places" (Gen 1:17; 2:8). He "blesses" (Gen 1:22, 28; 2:3). He "gives" (Gen 1:29, 30). He "completes" (Gen 2:2). He "sends" (Gen 2:5). He "breathes" (Gen 2:7). He "forms" (Gen 2:7, 19). He "plants" (Gen 2:8). He "causes" (Gen 2:9, 21). He "commands" (Gen 2:16). He "brings" (Gen 2:19, 22). He "takes" (Gen 2:21). He "closes" (Gen 2:21). He "fashions" (Gen 2:22). He "rests" (Gen 2:3).

The Creator initiates emergent and generative processes that anticipate a continuous creation with (in philosophical terms) "secondary causes." God desires a world that will in some sense be free from his direct control, and the creation is in certain ways marked from the very first moment by the presence of freedom. Freedom, we might say, is precisely one of the things that God creates. But the only way that freedom can be *created* is by opening a space in which freedom occurs or unfolds. The key refrain *Let*—"Let there be," "Let the waters," "Let the earth"—should serve as a clarion signal that God's way of bringing order out of chaos involves not only directly fashioning or controlling but also granting, permitting and delegating. We must think of the creation not only in terms of divine action but also in terms of divine restraint. Rather than simply dominating the world, in the very act of bringing the world into existence God is in a certain sense already withdrawing himself from it—or perhaps better, limiting himself within it—in order for it to be free. God is the sustaining ground of all being so nothing exists apart from God, yet the very fact that things exist that are ontologically other than God implies a simultaneously present/absent Creator from the very start.

Enter Adam. Like the animals before him, Adam is not created out of nothing nor by divine fiat but from preexisting materials taken directly out of the earth, which God "forms." The name "Adam," practically all commentators note, is etymologically related to *'adamah,* meaning "soil." Adam's name can be read to mean "from the soil" or even "the earth creature." Strangely, some believers have thought that the idea that humans might be related to other animals detracts from their glory as creatures uniquely made in the image of God. Being related to soil hardly seems like a more noble distinction. Yet the fact that humans share the same material origins as other animals is plainly stated in Genesis. In Genesis 2:19, we learn that God "formed every beast of the field, and every fowl of the air" from "out of the ground." Humans, birds and the beasts of the field thus share a common ancestry—they all come from out of the same earth. "Darwin and Feuerbach themselves could not speak any more strongly," writes Dietrich

Bonhoeffer. "Man's origin is in a piece of earth. His bond with the earth belongs to his essential being."[1] The language of Genesis is at once shrouded in tantalizing mystery and absolutely clear: Adam and the other animals are beings of the same matter, the same essential "stuff." They are intimately related, although the link that binds them is not simply matter but the forming hand of God. We do not know how God "forms" creatures out of preexisting matter. It would be a fatal mistake to imagine the Creator of the universe in anthropomorphic terms as forming with physical hands. But the intimate language of "forming"—of creating through the gathering and shaping of already existing elements—must be read as something very different from God's speaking things into existence.

As in the case of the animals, Adam is commanded to be fruitful and multiply. But Adam is given an additional task that may be our first significant clue to what Karl Barth referred to as the "shadow side" of the creation, the *nihil* or nothingness that a world created out of chaos and nothingness is always in danger of slipping back into.[2] In Genesis 1:28, Adam is told to *kabash* or "subdue" the earth. Elsewhere in the Hebrew Bible, the word *kabash* is used almost exclusively to refer to one thing: violent military conquest (cf. Josh 18:1; Judg 8:28; 2 Sam 22:40). The implication, then, is that there is a difficult and even martial task for humanity in relation to the rest of the creation. Adam must literally "dominate," "subjugate" or "conquer" the rest of the creation.

This sense is amplified in the account of the creation of Eve in chapter 2. In Genesis 2:18, God says he will make Adam an *ezer kenegdo.* "Helper" is too weak a translation, Robert Alter notes. What the word really connotes is an active sustainer or ally in military contexts.[3] We are therefore faced with a great riddle. Why is Adam charged by God with "subduing" (together with Eve, his strategic ally) the rest of the creation? According to some readers, such as seventeenth-century English philosopher John Locke, God gave Adam warrant to exploit or "own" the creation by adding his labor to it.[4] Genesis is thus read as a manifesto for capitalist appropriation, surplus theory of value and private property rights. Other commentators have tried to soften the language of subduing by reframing it in terms of "stewardship." But it may be that the language of military struggle actually serves a vital theological function.

As unsettling as it may be for some readers to discover, nowhere in Genesis is the creation described as "perfect." God declares his work to be "good" or *tob* at each stage and finally "very good"—*tob me'od*—at its end. Elsewhere in the Hebrew Bible *tob me'od* describes qualities of beauty, worthiness or fitness for a purpose but never absolute moral or ontological perfection. Rebekah is *tob me'od* or "very beautiful" (Gen 24:16 NASB). The Promised Land is *tob me'od* or "exceedingly good," its fierce inhabitants and wild animals notwithstanding (Num 14:7 NASB). When Joseph's brothers sell him into slavery the result is great hardship and pain for Joseph over many years, yet he declares that God providentially "meant it for *tob* in order to bring about this present result, to preserve many people alive" (Gen 50:20 NASB). According to the book of Ecclesiastes, "every man who eats and drinks sees *tob* in all his labor—it is the gift of God" (Eccles 3:13 NASB). In Lamentations, the prophet asserts that "It is *tob* for a man, that he should bear the yoke in his youth" (Lam 3:27 NASB).

In fact, Mark Whorton writes, nowhere else in Hebrew Scripture is *tob* or *tob me'od* interpreted by biblical scholars "as absolute perfection other than Genesis 1:31, and in that case it is for sentimental rather than exegetical reasons."[5] There are other words in biblical Hebrew that are closer to the English sense of "perfect" than *tob me'od* and that might have been used instead. The book of Leviticus commands that burnt sacrifices be *tamim*, "without defect" (e.g., Lev 1:3, 10; 3:1; 4:3; 5:18; 14:10). Elsewhere in Genesis, Noah is said to be *tamim* or "blameless" (Gen 6:9). In Deuteronomy 32:4, we read that God's "work is *tamim*" or "perfect," for "Is not He the Father who created you, fashioned you and made you endure!" (Deut 32:6 NJPS). Even in these texts, however, the biblical understanding of perfection or blamelessness lends little support to modern creationists. When we read Deuteronomy 32:4 in its full literary context, for example, we find that God's *tamim* work of creation—his "fashioning" of the children of Israel—

is revealed precisely in the long, perilous and conflictive process by which human civilizations evolved and the Israelites were brought out of "an empty howling waste" into a land of their own:

> The Rock!—His deeds are perfect,
> Yea, all His ways are just;
> A faithful God, never false,
> True and upright is He.
> Children unworthy of Him—
> That crooked, perverse generation—
> Their baseness has played Him false.
> Do you thus requite the Lord,
> O dull and witless people?
> Is not He the Father who created you,
> Fashioned you and made you endure!
> Remember the days of old,
> Consider the years of ages past;
> Ask your father, he will inform you,
> Your elders, they will tell you:
> When the Most High gave nations their homes
> And set the divisions of man,
> He fixed the boundaries of peoples
> In relation to Israel's numbers.
> For the Lord's portion is His people,
> Jacob His own allotment.
> He found him in a desert region,
> In an empty howling waste.
> He engirded him, watched over him,
> Guarded him as the pupil of His eye.
> Like an eagle who rouses his nestlings,
> Gliding down to his young,
> So did He spread His wings and take him,

Bear him along on His pinions;
The Lord alone did guide him,
No alien god at His side.
He set him atop the highlands,
To feast on the yield of the earth;
He fed him honey from the crag,
And oil from the flinty rock,
Curd of kine and milk of flocks;
With the best of lambs,
And rams of Bashan, and he-goats;
With the very finest wheat—
And foaming grape-blood was your drink. (Deut 32:4-14 NJPS)

If the reading I have offered so far is at all correct and God recruits the creation at each stage to play an active, participatory role in what follows, with Adam being charged with an especially vital task of "subduing" other parts of the earth, then there is a very good theological reason why God declares the creation to be "very good" rather than "perfect." The creation cannot be perfect because, in an important sense, it is not entirely God's work. There are principles of freedom at work in the creation, and animals, humans and the earth itself have a God-given role to play as his coworkers. There is also a strong sense that while the creation is in one sense "complete" at the end of the narrative, it is not yet *finished*. God "ended his work which he had made" (Gen 2:2)—that is, he completed what he had completed. But the story of God's creative purposes for his world has in fact just begun.

When God tells Adam to cultivate the Garden it is thus entirely consistent with the language and narrative arc of the story to see this cultivation as including the idea of expansion or development—God wants Adam to increase the Garden, and there is a tension between the world inside the Garden and the world outside it. "To be a creature is necessarily to be incomplete, unfinished, imperfect," writes Andrew Linzey in *Animal Theology*. "From this standpoint the very nature of creation is always ambiguous; it points both ways; it affirms and denies God at one and the same time. Affirms God because God loves and cares for it but it also necessarily denies God because it is not divine." Hence, "the state of nature can in no way be an unambiguous referent to what God wills or plans for creation."[6] The fact that God "rested" or "ceased" from his work on the seventh day may therefore represent not a termination point but a deeply pregnant pause. There is more to come, and we must wait to hear God say the words "It is finished."

Yet wherever there is true freedom at work there is also the possibility of deviation from, if not rebellion against, the divine will. For beings with moral agency and spiritual cognizance such as Adam and Eve, deviation would be rebellion, pure and simple. But we must ask whether beings possessing creaturely agency but not moral awareness might deviate in other ways that were simultaneously not in harmony with God's final plans but at the same time still within the sphere of what God would call "very good." By way of analogy, a wild horse must be "broken" by a skilled trainer. The wildness of the horse is part of its glory. It is *very good*. But its energies must be disciplined and channeled in new directions before it can become a horse that wins the greater glory, both for itself and for its master, of the Kentucky Derby.

In Genesis 1, there are implicit and explicit distinctions made between domestic and wild animals, or cattle and "beast[s] of the earth" as well as "creeping thing[s]." The "waters brought forth abundantly" "every living creature that moveth." The seas teem not only "with swarms of living creatures" (Gen 1:20 NASB) but also with "great sea monsters" (Gen 1:21 NASB)—not at all safe or domesticated creatures, but formidable predators, it would seem from the book of Job's vivid descriptions of the Behemoth and Leviathan. So there is a still-untamed and wild aspect to the creation. Adam and Eve must wrestle with this side of the created world and bring it more completely under God's dominion without overriding or exploiting its freedom. This is their high calling, and it may be a formidable task. The language of "subduing" in Genesis does not suggest pruning hedges. It suggests doing battle. Put another way, Adam's role is not simply that of a caretaker but that of a *redeemer*. The pressing question is: Might this wildness in the creation that still needs to be "subdued" or redeemed, emerging from principles of freedom or indeterminancy built into the creation, have included death as well? Could God have ever looked at a world that included death or pain of any kind and pronounced it "very good"? And could an untamed and very good creation have included elements of ferocity and even predation?

I will return to the question of animal predation in part two of this book. For the moment we must simply leave open the possibility that natural cycles of life and death involving plants and animals were indeed included in God's good creation in the beginning. The strongest argument against such a reading is Genesis 1:30, in which the animals are given "every green herb for meat." But we must note that while this verse hints against predation being *willed* by God, it does not resolve the question of whether the "great sea monsters" or other wild and creeping creatures might not at their first appearing in fact *be* predatory. Intriguingly, God does not address the animals prior to the creation of humanity with the words, "I have given you every green herb for meat," even though earlier in the text God does speak directly to the animals, telling them to "Be fruitful, and multiply." Rather, God (1) instructs Adam and Eve to "subdue" and "have dominion" over "every living thing" (Gen 1:28); (2) tells the human pair that to them he has given "every herb bearing seed" and "fruit of a tree" for "meat" (Gen 1:29); and (3) declares—still addressing Adam and Eve, the animals uniquely created in God's own image—that he has also "given" the other creatures "every green herb for meat" (Gen 1:30). It remains an entirely open question whether very good creatures lacking in moral awareness but possessing creaturely freedom or agency might not take that which they have not been given. Nor does the chapter speak directly to the question of *mortality* as such (predation and mortality being related but distinct matters, with the latter not requiring the former). Such an idea—that the creation was originally neither immortal nor placid but very good, mortal, finite and free—does not conflict with any clear verse in Genesis or the rest of Hebrew Scripture.

The idea that all death in nature is "cruel," "vicious," "sinful" or "evil" does not originate in ancient Jewish thought. As Walter Brueggemann writes, the Hebrew Bible does not assume any sort of "mechanistic connection of sin and death."[7] Death in classical Jewish thought is at times seen as a divine punishment or the consequence of sin, but at other times it is seen as the God-appointed fulfillment of full lives. One of the oldest Jewish commentaries on the creation narratives, the *Genesis Rabba,* compiled around the fifth century from still more ancient sources, declares that the divine assessment of "very good" or *tob me'od* in Genesis 1 is a kind of wordplay that implies *tob mot*: good death. God alone is able to see *all* of the creation, these experts of the Torah maintained, hence God alone is also able to declare the creation "very good," its provisional sufferings and death included, which to finite human beings with limited perspectives might at times appear as unmitigated evils.[8]

Whether or not we accept this ancient Jewish midrash, we must see that Genesis is in fact completely silent on the question of whether or not part of the wildness and freedom of the creation included mortality before Adam's fall, verse 30 notwithstanding. Yet this very silence may itself be evidence that death and even predation at least of a kind were indeed accepted in earliest biblical thinking about the creation process/event. If predators like lions and eagles were *not* part of the creation and the fact of their existence in the present was perceived as a theological problem by the Hebrew writers, we would expect Genesis or later books to provide some clues as to their origins or transformation, whether by divine "curse" or by demonical re-engineering of nature on a massive scale. Yet neither of these ideas is remotely stated in Genesis or in the rest of Jewish Scripture. Indeed, they are plainly contradicted by passages such as Psalm 104 (quoted at the start of this book) and the final chapters of the book of Job (which I will discuss in due course). Arguments from silence must, of course, be highly tentative. But the notion that all mortality and all predation in nature is the result of a divine curse is itself an interpolation, we must see, that has been imposed by pious readers on the great silences of the text. The creation narratives in Genesis are filled with lacunae and unanswerable riddles that should prevent careful readers from making very many dogmatic statements of any kind.

The only creature that is altered or "cursed" by Adam's fall is the serpent that was directly responsible for it, though its curse is not becoming a predator but rather being forced to crawl on its belly and "eat" the dust of the ground (Gen 3:14)—clearly metaphorical and symbolic language not to be taken literally. Some translations render the curse upon the serpent in Genesis 3:14 "*more* cursed," leading many readers to assume a general curse upon all animals. But the King James Version with its "cursed above all cattle, and above every beast of the field" comes closer to the actual meaning. The proper sense of the verse in Hebrew, commentators have long pointed out, is not comparative but rather selective.[9] The serpent is singled out from *among* the animals. Alter thus renders the passage, "And the Lord God said to the serpent, 'Because you have done this, cursed be you of all cattle and all beasts of the field.'"[10] To construe this simple statement to mean that God abruptly and supernaturally transformed docile creatures at every level of animal existence—not only in their instincts but also in their physical structures—into ferocious predators (or permitted a satanic being to do the same) is to take no small liberty with the text. Nor is there any mention in Genesis or any other book of the Hebrew Bible of mortality being imposed for the first time upon the nonhuman animal world as a result of human rebellion.

In Genesis 2, the reality of death comes directly into the foreground. Adam and Eve are placed inside a "Garden"—a place of fertility and safety—that is spatially and qualitatively set apart from the rest of the creation. The Garden is a sheltered environment while the world beyond—the world from which Adam was formed from out of the ground—is inhospitable if not hostile territory. Where Adam originally came from will, ironically, become the place of his banishment. East of Eden is his original home—that is, the "ground from whence he was taken" (Gen 3:23). When Adam is expelled from Eden it is therefore simultaneously an exile and a return.

There is a tree in the Garden that contains knowledge of good and evil. Adam is told by God that if he eats its fruit he will "surely die" (Gen 2:17). In order for the divine command to have gravity for Adam (no less than for us), he must understand what death *is*. Otherwise, God's words would for him would be the semantic equivalent of "you will surely X." Is the command intelligible to Adam because he has observed death elsewhere in the creation, perhaps in the land East of Eden "from whence he was taken"? Or has God revealed death to Adam, in the absence of any actual death, in some other way? We do not know. But the shadow of death, as an understood potentiality if not an observed fact of nature, is already a palpable presence in the Garden. Thus, when Adam and Eve eat the forbidden fruit their eyes are opened *not* to death, according to the text, but to their own nakedness, to their vulnerability and dependence. The result of human rebellion, careful readers must also note, is not the miraculous invention of nerve endings in humans capable of experiencing pain for the first time where no pain of any kind previously existed. Rather, we are told, Eve's pain in childbirth would be greatly *multiplied* (Gen 3:16). Humans, Genesis plainly suggests, knew pain before the fall even if not prolonged suffering.

The first unmistakable death is recorded in Genesis 3:21—and it is by all indications God who is responsible for it. "Unto Adam also and to his wife did the Lord God make coats of skins, and clothed them." Why skin and not wool? The traditional answer is that God's act of animal slaughter is a form of moral pedagogy. Adam and Eve must learn that the consequences of their sin are death. But an equally plausible (and perhaps not mutually exclusive) reading would be that God knows that Adam and Eve's coverings of fig leaves will not be enough to protect them in the unsheltered, untamed world *outside the Garden*. He mercifully gives them the kind of clothing they will need for their new environment. Readers who hold to a high view of Scripture's authority must be very careful, then, about projecting their own notions of perfection and goodness onto the text and onto nature in the name of defending God's character. These readings may in fact pose far greater theological and moral perils than the idea of death of a kind before

Adam's rebellion. To say that seasonal changes and cycles of birth, life and death in nature are, without qualification, "satanic," "evil" and things we will someday escape by leaving this veil of tears and illusions behind, may actually be an expression, in Jewish perspective, of ingratitude if not contempt for God's good creation and the earthiness of material existence. Indeed, Alan Lewis notes in *Between Cross and Resurrection*—his meditation on the Christian meaning of the sabbath written as he entered the final stages of terminal cancer—that the very nature of the human rebellion may lie precisely in Adam and Eve's refusal to accept their own nondeity, the temporality and finitude of their creaturely status.[11]

Such a reading of Genesis clearly opens the door to the possibility of "theistic evolution"—or better, *process creation*—with believers seeing evidences of redeeming purpose at work in a universe that is neither wholly determined nor wholly contingent where Darwin saw only laws of ironclad necessity (competition) and chaotic chance (random mutations). "Not many Christians today find it necessary to defend the concept of a literal six-day creation," writes Anglican evangelical clergyman John Stott, "for the text does not demand it, and scientific discovery appears to contradict it. The biblical text presents itself not as a scientific treatise but as a highly stylized literary statement." Stott continues:

> It is most unfortunate that some who debate this issue (evolution) begin by assuming that the words "creation" and "evolution" are mutually exclusive. If everything has come into existence through evolution, they say, then biblical creation has been disproved, whereas if God has created all things, then evolution must be false. It is, rather, this naïve alternative which is false. It presupposes a very narrow definition of the two terms, both of which in fact have a wide range of meanings, and both of which are being freshly discussed today.[12]

It would be a form of cultural imperialism, John Walton concludes, to read Genesis as a scientific or historical record.[13] It is a theological text concerned with theological meanings, and it has conveyed these meanings in a form that must be grasped in terms of the prescientific worldview of its original hearers rather than our own modern one.

Yet this theological vision *does* impose important limits on other historical narratives of origins and on the scientific worldview. Genesis is not a story of material origins, Walton argues, but of "temple inauguration" in which God assigns the different parts of the cosmos their ordered *functions* in six days, however long it may have taken for them to be materially formed. God then takes up his residence in this cosmic temple on the seventh day. Self-described literalists, Walton contends, have failed to understand what the Hebrew word "create" (*bara'*) literally means, although scientists who stridently declare that there is no purpose in nature also go far beyond what the material evidence can tell us. Whether or not one accepts Walton's (to this reader's mind compelling) argument, Genesis reveals that material and biological existence was and is created by a loving God, even if through a material process we do not fully understand. And no matter how much contingency we may observe in nature, we might still discern an underlying order and sacramental mystery in its fine-tuning—a deep resonance, as Alister McGrath writes, with the theological concerns of the biblical writers.[14] Genesis tells us that humans are intimately related to this world and to other creatures, but that we are also qualitatively related to God in a unique way that includes dimensions of moral awareness, reasoning and responsibility that cannot be grasped in reductive or purely materialistic terms. It tells us that the creation has been marred or distorted as a result of human rebellion. And it reveals that we are now alienated from each other and from God as a consequence of our fallen natures.

2

What's Eating Biblical Literalists?

Creationism and the Enlightenment Project

❧

My reading of Genesis has taken the form it has because I have been writing from the start with questions of modern science and evolutionary biology very much in mind. My goal here is not, however, to reconcile Genesis with Darwinian theory. It is to help clear a space in which questions about the relationship between the biblical vision (or visions[1]) of creation and the evidences of modern biology and geology can be asked without fear, rancor or dogmatism—whether of the religious *or* the scientific kinds. My interpretation of Genesis—the reflections not of a biblical scholar but of a lay theologian concerned with locating a more ethical hermeneutic or hermeneutical ethic in the face of moral challenges that confront all believers—is strictly textual and in no sense dependent on modern scientific models or historical-critical methods. At the same time, I am open to whatever genuine knowledge about physical realities and history can be gleaned from the weight of the scientific and historical evidence. I have tried to read Genesis literally in order to understand it *theologically* but without succumbing to a wooden literal*ism,* biblicism or fundamentalism. I would happily describe myself as a literalist and a creationist in my thinking on Genesis, then, if others were to grant me the titles. I have done my best, in any case, to try to understand what Genesis is literally concerned with, and I believe that God is the Creator the universe and the giver of life. I realize, however, that this is not what the terms "literalism" and "creationism" are now widely understood to mean, and so to avoid confusion I will (with no small reluctance) follow the common usage.

Biblical literalism and modern creationism—what would more accurately be called *concordism*—are approaches to Genesis that insist, among other things, on the scientific and historical harmony (or "concord") of the primeval stories (Genesis 1–11) as defined by contemporary notions of scientific and historical objectivity, regardless of the actual weight of scientific and historical evidence. Old Testament scholar Gerhard F. Hasel succinctly describes and defends this approach to the Bible as follows: "whenever biblical information impinges on matters of history, [the] age of the earth, origins, etc., the data observed must be interpreted and reconstructed in view of this superior divine revelation which is supremely embodied in the Bible."[2]

There are, of course, parts of Scripture that all scholars believe should be read as historical, and there are parts of Scripture that even the most literal-minded readers agree should be read in nonliteralistic ways. For the purposes of this book I will use the term "literalism" to describe any approach to the Bible that treats the creation story in Genesis 1 as a chronological-historical account that should be used to coordinate and explain scientific and historical data—the enterprise of "creation science" or "scientific" creationism.[i] But why, we must ask, should we assume that not only theological but also scientific and historical truth is uniformly or "supremely" embodied (as Hasel claims) in the biblical canon or in the creation stories of Genesis in particular?

---

[i] I have placed these terms in scare quotes throughout this book to signal the fact that we are dealing with a research project that might on occasion make use of properly scientific methods but that is not, in the final analysis, bound by the rules of evidential and inductive reasoning normally associated with scientific approaches. "Scientific" creationism, as Hasel's statement makes clear, cannot be scientifically falsified or challenged in any way by the actual weight of the empirical evidence.

Many Christians strenuously resist the idea that their readings of the Bible are driven as much by philosophical as by theological commitments. Merold Westphal describes an advertisement he once saw for a new translation of the Bible being billed by its publishers under the banner, "NO INTERPRETATION NEEDED."[3] Yet the notion that the Bible ever speaks so unequivocally or with such immediacy that no interpretation is needed, Westphal points out, is itself an interpretation of an interpretation with a highly problematic pedigree. This philosophical school of thought—the school of "naive" or "commonsense" realism—is itself in need of interpreting since what is "plain" or "commonsense" to one person might not be obvious or uncontroversial at all to another (whether because they possess more or less information, or because they simply see the same facts in a different light). We all bring important background experiences and beliefs about the structure of reality with us to our readings of the text, and this means there are no "plain" or purely "religious" readings of Scripture untainted by philosophical perspectives or by our culturally embedded worldviews. The most urgent questions facing creationists in this light are not textual but rather epistemological. They are, we might say, *prehermeneutical.*

Biblical literalism as a self-conscious epistemological stance, Nancey Murphy in particular has helped us to see, is an expression of the philosophical foundationalism and rationalism that underwrites the Enlightenment project and modernist worldview.[4] Some care in terminology is needed here. We are all in a certain sense "soft" foundationalists insofar as we all have basic or "foundational" assumptions about the nature of reality, with some of our beliefs being more important than others. My belief in the triune character of God and in the incarnation of Christ is more foundational to my life than my belief that *The Godfather* is a great film. But this is not what the term *foundationalism* means to philosophers. First and foremost, "scientific" creationism (insofar as one may generalize about what is in fact a diverse and often deeply divided movement) reflects a relatively recent notion. This notion is shared, ironically though not coincidentally, by many thoroughgoing naturalists—that all truth claims must be stacked one atop another like the bricks to a house, beginning from an indubitable "firm foundation." This idea has roots stretching all the way back to Plato but took a new and greatly intensified hold upon Western thought around the middle of the seventeenth century largely as a result of the writings of René Descartes.

In his *Discourse on Method* published in 1637, Descartes declared that the search for truth must begin with the deconstructive task of clearing away all uncritically held beliefs—including even beliefs derived from one's direct observations and sensory perceptions. One should systematically doubt every idea and every truth claim that cannot produce irrefutable, geometry-like formal proofs of its validity. Only after one is able to locate a source of absolute certainty can one begin the reconstructive task of establishing knowledge through a chain of demonstrations or proofs building upon one's infallible first principles.

After isolating himself from society and pursing a series of intense thought experiments in systematic doubt, Descartes concluded that the one thing he could not objectively deny by following the path of radical skepticism was the fact of his own subjectivity. The base of Cartesian certainty is the inescapable fact of the doubter's own doubting, that is, his or her mental processes. In Descartes's famous dictum, *Cogito ergo sum*—"I think, therefore I am":

> But immediately I noticed that while I was trying thus to think everything false, it was necessary that I, who was thinking this, was something. And observing that this truth, "*I am thinking, therefore I exist*" was so firm and sure that all the most extravagant suppositions of the skeptics were incapable of shaking it, I decided that I could accept it without scruple as the first principle of the philosophy I was seeking.[5]

The subjectivism of the Cartesian method, though, in many ways amounted to a denial of the subjective realities of felt and lived human experience. In opposition to Renaissance humanists such as Thomas More and Desiderius Erasmus (who, unlike the "secular humanists" of present-day fundamentalist obsession, were deeply committed Christians), Descartes sharply distinguished between reason and emotion as well as between bodily and mental realities. He disparaged the particular, the local and the timely in the name of the universal, the general and the timeless. He sought to reframe all questions of human knowing in terms of a strict rationalism, casting aside the earlier humanist embrace of a pliant and tolerant pluralism in the search for truth. The older religious traditions were rooted in recognition of the fundamentally ambiguous nature of much of existence and the need for epistemological humility as a perhaps even higher virtue than epistemological certitude. But certitude, not humility, was what Descartes was after, and so what could not be assimilated within his system of skepticism/rationalism was shunted aside. The transition from medieval philosophy and Renaissance humanism to the modern scientific worldview with its Cartesian oscillations between radical doubt and colonizing regimes of dogmatic, universalizing certainty, Stephen Toulmin writes, was marked in significant and disturbing ways by "a narrowing in the focus of preoccupations, and a closing in of intellectual horizons."[6]

What does this have to do with biblical literalism and present-day creationism? Similar to the driving spirit of Cartesian rationalism, what we witness in the culture of "scientific" creationism is a dramatic narrowing in the focus of religious concerns, a closing in of intellectual horizons, a strict demarcating of the boundaries of thinkable thought. Cartesianism and creationism in fact bear a striking family resemblance, with creationist publications being replete with foundationalist vocabulary. Those who cast themselves in the role of heroic knights of faith waging battle against the dragons of secular modernity ironically replicate and perpetuate the very anxieties (and hubris) that underwrite much of the modern worldview. In the words of Carl Raschke, highly conservative theologians wedded to foundationalist reasoning "rhetorically maintain God's unshakable power and presence, [but] they do so by following modern philosophy to midnight worship on the high places."[7] Many do so, we might add, as "sleepwalkers" who lack awareness of the historical and philosophical sources of some of their most deeply held religious convictions.

Creationists have posited a very different source of absolute certainty to the Cartesian self-grounding ego, to be sure. For biblical literalists, the firm foundation or base of universal and infallible knowledge on which all truth is said to rest is a "plain" reading of Scripture alone, with particular emphasis on the first verses of the book of Genesis. According to Henry Morris:

> If God did not mean what He said in the very first chapter of His book, then why should we take the rest of it seriously? It seems to us that we should take Scripture as the literal Word of God, intended to be understood by its readers in every generation and every nation—especially *this* chapter [Genesis 1], which is the foundation of all the rest.[8]

Biblical texts, with varying degrees of sophistication, are thus presented in creationist literature as "proofs" in chains of apologetic argumentation that all add up to a single fearful claim: *The beliefs that have long given us our sense of certainty, meaning and hope will all come unraveled if we read Genesis as anything other than a precise historical account of origins, just as we have always done.*

"I would rather die than believe in evolution," I once heard a creationist proclaim. He no doubt meant to impress upon his listeners the courage and strength of his convictions, although I was struck at the time by how insecure and unhappy he sounded. If young earth creationism is wrong, another highly educated man writes, "Christianity would be falsified." So what would this scientist do if God himself sent him a vision telling him he was mistaken in his assumptions? Would he humbly acknowledge that not Christianity but his creationism had been falsified? Or would he dismiss the vision as a diabolical test of his faith to be strenuously rejected because it did not fit with his prior beliefs? Creationists are no doubt right that we should not believe *in* evolution—just as we should not believe *in* the Great Depression or believe *in* Gandhi's great Salt March (or any other historical process or event). The strange syntax of belief *in* evolution aside, we should base our beliefs *about* natural history no less than human history on the weight of the evidence, remaining very open to where the evidence might lead. Many literalists, though, live with a visceral terror, thinly veiled behind their statements of dogmatic certainty and superior faith, that the entire religious edifice they have dedicated their lives to constructing could at any moment come crashing down upon their heads. Theirs is a theology conceived as a high-stakes game of Jenga. Whatever you do, *don't touch the bricks at the base of the tower*.

The foundational importance of creationism for all Christian belief and practice is allegedly self-evident from the objective words of Scripture, so that strict literalism on Genesis cannot be *subjectively* denied by anyone who truly has faith in the Bible's authority and has read Scripture with intense inner devotion under the guidance of the Holy Spirit. All external sources of knowledge, including even one's direct observations of empirical realities, must be regarded with an attitude of skepticism and doubt until situated in the reconstructed tower of knowledge built upon these putatively incontrovertible biblical foundations. Literalism and young earth or young life creationism are therefore varieties (although some creationists may protest otherwise) of the theological and epistemological stance known as *fideism*.[9] They rest upon the conviction that human reason left to its own ways is not merely inadequate to arrive at full theological knowledge but is in some sense antithetical or *hostile* to faith. Human reason leads the mind disastrously astray from spiritual as well as empirical truths until/unless radically subordinated to a simplifying and absolute religious authority through an act of intellectual submission and humble obedience of the will (although we were not especially humble, as I recall, when we used to sing in Sabbath School, "God said it, I believe it, and that settles it for me!").

Yet we must note what is actually most foundational in the literalist's fideist stance. It is *not* the authority of Scripture, which nonliteralists also fully accept, but rather belief in philosophical foundationalism itself as the unquestionable stage on which all hermeneutical battles must be played out.

Biblical literalists do not approach the text with a *tabula rasa*. Like all people, they have a worldview and philosophical assumptions that are antecedent to their readings and so powerfully shape their conclusions. And their worldview so happens to be deeply implicated in the modernist worldview that Descartes and other Enlightenment thinkers helped to create. Literalists are convinced that the only way to have truth of any kind is through the sort of mental bricklaying urged by Descartes that treats different sources of knowledge in a divided and hierarchical way, and this leads them to a picture of the life of the mind as an essentially one-way street. One's "base" of epistemological certitude must have a profound influence on the secondary "floors" of knowledge one constructs, but these floors cannot alter one's base in turn. Creationism is in this sense not merely indubitable. It is *incorrigible*—it cannot be undermined or altered by any new information, whether arriving from "above" or "below." It is impervious to the weight of empirical and historical evidence. Scripture, as the self-validating ground of its own internal truth claims, must change our views of scientific evidence, but new scientific and historical evidence cannot reciprocally challenge or reshape our reading of Genesis in any meaningful sense.

Individuals such as Morris and Hasel—no different than strident atheists such as Richard Dawkins and Daniel Dennett—in this way deny the possibility of epistemological and methodological holism in favor of a ranked ordering of epistemic values. It is telling that many creationists in fact cite writers such as Dawkins and Dennett with approval to support their claim that we are faced with only two live options: fundamentalist-style creationism or atheistic Darwinism. Creationists and ultra-Darwinians (as they have sometimes been called) actually stand united in their pictures of reality, religious faith and the life of the mind at a much deeper level than their highly conflictive and mutually exclusive declarations might lead one to believe. Both groups uncritically share the same essential dualisms: religion *versus* ("atheistic") science, faith *versus* (natural) reason, special revelation *versus* (normal) human experience and observation, divine fiat *versus* (purposeless) evolution. At the same time, the

Bible's authority for biblical literalists awkwardly hinges precisely upon its alleged *scientific veracity*. Literalists paradoxically reject the findings of modern science in the name of Genesis' superior scientific reliability. Strict literalism on the creation is therefore a rejection of modern science by individuals who have already drunken deeply (even if unconsciously) from scientism's wells.

We can now make better sense of Hasel's claim that believers who read even the seemingly obscure genealogies in the first chapters of Genesis as theological or symbolic or literary passages instead of straightforward historical records betray their "accommodation to the physical and life sciences" and "acceptance of modernistic and/or evolutionary patterns of the origin and history of our planet and life thereon."[10] We must be prepared to deny the overwhelming weight of scientific and archaeological evidence if need be, Hasel makes clear, in order to defend the supreme scientific and archaeological authority of Scripture. This includes rejecting the overwhelming consensus of historical, scientific and archaeological scholarship showing that there were settled communities in China by 5000 B.C.E. and in the Near East by 7000 B.C.E., not to mention tool-making cultures in the Americas by 10,000 B.C.E. (if not earlier) and exquisite Paleolithic cave paintings in Chauvet, France, dating to more than 30,000 years ago. All archaeological discoveries must somehow be dated, if I read Hasel correctly, to *after* 3402 B.C.E.—the oldest proximate date in his calculations (from the Greek Septuagint) for Noah's catastrophic flood, which is said to have completely erased all prior human civilization, violently divided the continents, carved the Grand Canyon, thrust up the Himalayas, and so on.[ii] No merely localized Near Eastern or

---

[ii]Although 3402 B.C.E. or very close to it is the oldest date for Noah's flood that Hasel seems willing to accept, he does not rule out the traditional Masoretic text in Hebrew, which includes chronological variations from the Septuagint and generates a flood date closer to 2400 B.C.E.—by which time, problematically, the first pyramids of Egypt had already been built. Other literalists are less rigid when it comes to questions of chronology, allowing for a flood thousands of years earlier than this to better accommodate archaeological data (without explaining why it is in principle acceptable to take archaeological data into account when interpreting Scripture but not biological data).

Black Sea flood as has been hypothesized by some scientists will do.

The days of the creation week in the first chapter of Genesis, it goes without saying, can likewise only possibly refer to "literal" chronological or stopwatch time. For literalists, Genesis 1 and 2 narrate events in approximately the same way the BBC or CNN would have captured them had their reporters been present with television cameras rolling. There is one way to read Genesis, they insist, and they charge all others with doing violence to the plain, self-evident meaning of God's Word, whose foundational authority would crumble were it not for the text's one-to-one correspondence with modern historical and scientific standards of truth.

3

Unwholesome Complexity

Literalism as Scientism's Pale Mimetic Rival

In support of these strong assertions about the meaning of Genesis, literalists have often relied on stock arguments incestuously circulated in creationist literature. Much ink has been spilled, for example, seeking to show that the Hebrew word for "day," *yom*, in Genesis 1 means "day" in the sense of a literal twenty-four-hour period rather than a symbolic eon or age as in allegorical readings (which are also essentially concordist in their approach). This is undoubtedly correct. Allegorical readings that suppose a one-to-one symbolical correspondence between the days of Genesis and historical epochs are entirely unconvincing. Few literalists would conclude, however, from the numerous references in Scripture to the "hand" of God—an equally literal word—that the Creator of the universe possesses a physical body or that the theological meanings of these passages require that God possess ten literal fingers. Pointing out that *yom* in Genesis semantically means "day" hardly settles the question of what kind of literature Genesis is, how it ought to function in the life of Christian faith, or what it is actually bearing witness to as far as scientific questions are concerned. This is to say nothing of the question of whether its meanings for readers today should be exactly the same as they might have been for its original hearers. Practically all scholars agree that the serpent in the Garden was not originally thought of in Jewish belief as the accuser angel of later Hebraic and Christian belief, although this evolution of theological meaning is entirely permissible; original interpretations, most believers accept, should not always be binding upon subsequent ones, and sometimes texts must be read "backward" from the perspective of new knowledge. The task of biblical theology is not simply to rethink the thoughts of the biblical writers after them without modification, if such a thing were even possible.

Then there is the widely quoted statement in creationist literature by James Barr, a distinguished Old Testament scholar who taught at Oxford University and passed away in 2006. In 1984, Barr wrote a letter in which he said:

> Probably so far as I know there is no professor of Hebrew or Old Testament at any world-class university who does not believe that the writer(s) of Gen. 1–11 intended to convey to their readers the idea that creation took place in a series of six days which were the same as the days of 24 hours we now experience.[1]

In his 2007 defense of strict literalism and young earth creationism, *The Six Days of Genesis: A Scientific Application of Genesis 1-11,* Paul Taylor quotes the letter with citation to a 2003 article by his cocreationist Jonathan Sarfati.[2] Sarfati in turn cites the letter but with improper citation in his endnote, making it impossible to know where exactly he found the document.[3] The letter is meanwhile quoted by John Morris in his 2007 creationist book, *The Young Earth,* with citation to a 1993 creationist article by one Russell Grigg (published by Creation Ministries International, which seems to be the primary disseminator of the Barr quote and whose magazine has confusingly undergone four name changes in the past twenty years, currently appearing under the masthead *Journal of Creation*). Grigg's footnote simply reads, "Copy held by the author."[4] And so the search for a primary source again runs cold. I have so far been unable to locate the original source of the Barr quote, because every time creationists cite him it seems to be with reference to the works of their fellow creationists or to a "Copy held by the author." The earliest published use of the Barr quote seems to be a 1986 book by John Whitcomb titled *The Early Earth*. Whitcomb's citation does not list a published source, however, but simply provides a date for the letter and the name of its addressee (David Watson). I do not mean to imply that Barr did not say what creationists say he did, which I have no reason to doubt. Yet Barr's statement, philosopher Alvin Plantinga shows, is demonstrably false.[5] It has also been used in highly misleading ways in the self-referential world of creationist apologetics.

Numerous Hebrew scholars, whose work Barr would have been well aware of, have long held divergent views on what Genesis meant to its original hearers. In his 1982 commentary on Genesis, Walter Brueggemann, to cite a single example, suggested that the creation narrative in Genesis 1 was probably composed for poetic and liturgical purposes as late as the sixth century B.C.E. in the context of the Babylonian exile. "[I]t has been urged that [the creation account] is a historically descriptive account of what 'happened,'" Brueggemann writes. "But that kind of scientific, descriptive reporting is alien to the text and to the world of the Bible."[6] Further, Barr—a liberal scholar—made the claim he did as part of a highly polemical attack upon evangelical Christians who did *not* subscribe to literalism on the days of Genesis. His goal was to show that self-described literalists pick and choose which parts of Scripture they will be literal about.

The fact that there is no uniformity of scholarly opinion therefore empties the Barr quote of its entire rhetorical force in creationist arguments, while the larger point he was trying to make also does not support but in fact undermines the claims of biblical literalism. Creationists have appealed to Barr to prove that even those who disagree with their science, if honest, cannot rationally disagree with their claims about the meaning of Genesis. But of course many committed believers with strong credentials as biblical scholars can and have disagreed. The Barr letter should thus receive a peaceful burial, not to be quoted from again unless as a warning tale of the dangers of two logical fallacies: the fallacy of the uncritical appeal to authority (on the part of creationists) and the fallacy of the hasty generalization (on the part of Barr).

The greatest problem with strict literalism's "plain" reading approach to Genesis, however, is that it is not nearly plain or literal enough. Creationists have treated Genesis as a story that is all surface with no depth that must now be validated or "proved" through—irony of ironies—the tools of a thoroughly rationalistic, quantifying and materialistic science. But the demand for scientific and historical correspondence—the criterion of "truth" demanded by modern, post-Enlightenment minds—introduces unwholesome new layers of complexity to our readings. These layers are not located *inside* the text, drawing us into its mysterious and undisclosed depths as I have attempted to do in my reading in chapter one, but rather are piled on top of the story from without, strangling its poetic and doxological heart. At one time, the cutting edge of "creation science" included the search for proof that the firmament of Genesis 1:6 was a literal hardened canopy of polarized hydrogen ice crystals. Today, self-described literalists on Genesis lead us through dense philological and semantic thickets in order to prove that the two distinct creation accounts in Genesis 1 and 2 can be perfectly harmonized according to external scientific and chronological criterion. (They posit, for example, that the plants created on day three of the creation in Genesis 1 did not include *agricultural* plants, to explain why vegetation in Genesis 2 is said to be created *after* the creation of Adam.[7])

Accepting *arguendo* the literalist claim that we can make the chronologies of the two creation accounts perfectly conform on purely linguistic grounds, such a "scientific" reading unfortunately cannot be sustained on *narrative* grounds without producing a picture of God's creative activity that is inadvertently comic and that greatly detracts from the beauty and theological seriousness of both creation stories when read on their own terms. I do not wish to make light of the beliefs of others and certainly not the words of Scripture. But when self-appointed guardians of the faith place senseless obstacles in the paths of honest seekers after truth, I see no reason not to name the folly.

Genesis 2, in contrast to Genesis 1 by any straightforward reading, paints a picture of Adam being created a significant time before the rest of the animals and even before the creation of plants. On his first coming into the world Adam thus finds himself situated in a barren if not harsh terrain. But the Creator will provide for the earth creature in his vulnerability and naked exposure. God creates the Garden of Eden and places Adam inside it (the Garden is not mentioned at all in chapter 1). Adam then spends a meaningful duration working in it alone. God decides that it is not good for Adam to be alone and so creates the animals, which he brings one by one or kind by kind to Adam to see what he will name them.

It is only after this process, involving a seemingly significant passage of time, that God causes Adam to fall asleep and creates Eve from one of Adam's ribs. Unlike the first creation narrative in Genesis in which humans are created together, male and female, as the climax of the creation after all the other animals have been formed, in Genesis 2 the creation of nonhuman life is bracketed or encircled by the creation of humanity, male and female, which comes first *and* last in the story. This is a very different although completely complementary *theological* statement of what it means to be made in the image of God and in relationship to the rest of the earth's creatures. Or so any plain reading of the story would lead one to conclude.

Here, however, is what we must picture happening in the daylight hours of day six of the creation by any reading that flattens Genesis 1 and 2 into a single linear historical narrative with an eye to scientific correspondence and strict chronological sequence rather than to things such as literary technique and complimentary theological meanings:

At the start of day six, God commands the earth to "bring forth the living creature after his kind," and the earth brings forth cattle, reptiles and wild animals (Gen 1:24). Only *after* the land animals have come forth from the earth (ignoring the plain implication of Gen 2:18-19) does God create Adam from out of the same dust of the ground and breathe into him the breath of life (Gen 2:7). God next creates the Garden (for somewhat inexplicable reasons considering the entire world is already a verdant, nonthreatening oasis in this account), placing Adam in it with instructions to till and to keep the Eden paradise and to avoid the fruit of the tree of knowledge of good and evil (Gen 2:15-17). Adam commences tilling the land but is, apparently, immediately filled with feelings of ennui and loneliness. God decides it is not good for just-created Adam to be alone (Gen 2:18). At once a massive stampede of animals (until then apparently hidden from Adam's sight as he worked the Garden) comes crawling, flapping and galloping past the no-doubt-bewildered man who only came into being hours (if not minutes or seconds) before. According to the text, the procession includes "every" beast of the field and bird of the air. Adam hastily names the creatures but no suitable partner is found for him (Gen 2:20). God at this point induces sleep (and it is a "deep" sleep, or "trance" in some translations, so not something that happens instantaneously, plain readers might infer, although "scientific" readers might postulate otherwise), removes a rib, creates woman, revives Adam and offers introductions (Gen 2:22-23).

Finally, we arrive at the *pons asinorum* of this scientific conflation of the two creation accounts. It is imperative to the life of faith, creationists admonish us, that we accept that all of the above occurred within twenty-four hours and not a minute more. "[W]e will never make room for anything other than a literal six-day creation for life here—never," writes one administrator in the flagship journal of the Seventh-day Adventist Church, the *Adventist Review*. "And for those who want more, you'll have to fight us for every extra minute—much less your millions of mythological years beyond—of which the Word of God knows nothing and with its first verses utterly denies."[8] The creation of life on earth in its entirety, Adventist officialdom today asserts with somewhat astonishing confidence in the incontestability if not the infallibility of its hermeneutical formulations, took place "in six literal, consecutive, contiguous, 24-hour days of recent origin," "identical in time to what we now experience as a week."[9]

Unfortunately, statements such as these—which fit within larger patterns of rising fundamentalist religiosity around the world—are not nearly specific enough in the light of the findings of modern physics. Never mind the fact that the earth's actual rotation at present is several minutes less than twenty-four hours and is slowing every year (the best scientific evidence suggests that two billion years ago a day on earth lasted about ten hours). Einstein's special theory of relativity states that time itself is elastic and may be shrunken or stretched—or even stopped altogether in what physicists refer to as a "singularity"—by motion and gravity. Contrary to all commonsense notions derived from daily experience, time actually runs faster where there is less gravity (a fact that has been empirically demonstrated by flying clocks in space). Because time—or more accurately, spacetime—is not constant across the universe but relative to motion and gravitational forces, we can no longer even speak in a simple way of the division of time into past, present and future. These categories may have meaning from our own particular vantage point, but in the mind-bending world of the new physics what is happening *now* on earth can be happening in the past or even the future from the perspective of observers moving at different speeds in other parts of the galaxy. A thousand years of stopwatch time on earth might be a single day for someone traveling very close to the speed of light, and if one could travel faster than the speed of light it would even be possible to travel backward in time.

Hence, those creationists keen to maintain doctrinal purity through the most excruciatingly precise language about the meanings of Genesis must now either deny the findings of modern physics no less than evolutionary biology, or else specify the exact location where they would have us position our clocks when counting the days of Genesis. They must make sure that we think of time in God's creation of the universe in strictly anthropocentric, geocentric, and Newtonian terms that perfectly coincide with our normal psychological and sensory experiences. Thus: "The Creation in all of its parts occurred in six literal, consecutive, contiguous, 24-hour days of recent origin, identical in time to what we now experience as a week, *as measured by a clock located at a fixed spot on the surface of the earth, GMT—Garden Mean Time.*"

If one were a mathematician one might at this point begin calculating how many seconds Adam had to name "every" beast of the field and bird of the sky during his frenetic first day of existence. (There are 1,440 minutes in a day and some 9,000 species of birds alone, not including the extinct ones, and we are told by creationists that there has been no significant evolution of new species since the creation event.) I am not a scientist or mathematician, however, so I will simply attend to the lesson we have gained by conflating Genesis 1 and 2 on "scientific" and chronological grounds rather than allowing them to speak to us as complementary theological texts that tell us about the creation in a highly allusive literary form never meant to be submitted to these kinds of thoroughly modernistic and rationalistic pressures. The most important new insight we have gained, it would seem, is that man cannot survive for more than a few hours without a woman—and that Einstein is potentially as dangerous for believers to read as Darwin.

But can one really be a biblical literalist holding to a firm "Thus saith the Lord" on both Genesis 1 and 2 without strain or contradiction? Or is the effect of being a literalist on chapter 1 that one must be evasive, equivocal and in the end violent with the language of chapter 2 in order to make the chronologies somehow "work"? One might say as some literalists have—picking and choosing what they will be literal about, just as Barr warned us literalism inevitably must do—that Adam didn't really name *all* of the beasts of the field and birds of the air but only a few representative samples, or perhaps only the ones in his particular vicinity of the Garden. Alternatively, we might equivocate on "beasts of the field" and try to restrict its meaning to a more manageable number of domestic animals (goats, donkeys, llamas, Indian peafowls, etc., even though in Genesis 3:1 it is clear that reptiles are included as well). In any case, we must somehow make Genesis 2 say something other than what the text very plainly appears to say, not because we are really concerned with listening to the story on its own terms as a theological narrative but because we have a prior commitment to an uncompromising and thoroughly modernist understanding of what counts as "truth," so that all other textual questions must now be subordinated to the task of producing absolute scientific and chronological harmony—no matter what the texts say.

Still another example of the unwholesomely scientistic way literalists have come to read their Bibles is the way they have pressed the story of Noah's flood into the service of "flood geology." The violence of the flood, creationists believe, was on a scale almost beyond comprehension. In the words of Harold W. Clark in his 1929 fundamentalist classic, *Back to Creationism*:

> The scattered and irregular distribution, the twisting and tilting, the stupendous carving and upheaval of the rock formations seen in any mountain region, point to violent action far surpassing anything that men know today. . . . [T]he wonder is not as to how the Flood could have happened, but as to how such violent action as we find indicated could ever have stopped short of completely wrecking the earth and destroying absolutely all life upon it.[10]

In their enthusiasm to use the story of Noah's flood to somehow coordinate with the data of modern geology, however, few creationists have noticed the way their literalism strains credulity for *biblical* reasons, precisely in the light of the heavy *scientific* lifting they want the story perform.

What should we make of the fact, for example, that the geological event that is said by creationists to have wreaked such havoc upon the earth's surface, literally tearing apart its mantle, somehow left all of the preflood geographical markers mentioned in Genesis 1–10 (the Tigris and Euphrates Rivers, the land of Cush, the land of Nod) precisely in place once the flood waters subsided? Did God miraculously preserve the landscape of the Near East from the forces of the flood so that Noah would be able to take his bearings when he disembarked from the ark? Did Noah's descendants nostalgically name new postflood rivers and territories with the same names as preflood ones (in which case, for all we know, the antediluvians might have inhabited present-day South America or Antarctica)? Or instead of such special pleading might it not be worth considering that the real concern of the Genesis writer (or writers) was theological rather than scientific in nature and that the story of Noah's flood was meant to tell us about God's providential care of people and animals in the face of human violence (Gen

6:11), even if nothing whatsoever about God's historical creation of the Matterhorn and Mount Kilimanjaro?

The tragic irony of strict literalism and "scientific" creationism, I am suggesting, is that it presents itself as the great opponent of scientism when it is in fact one of scientism's most acute manifestations. The kind of fideism that declares Scripture to be the highest guide to truth as over and against modern science while simultaneously insisting that the authority of Genesis is contingent upon its scientific veracity is, on closer examination, no fideism at all. It is, rather, scientism's reactionary doppelganger and pale mimetic rival, enraptured by the very thing it seeks to resist. The results are healthy neither for science nor for the life of faith.

Literalism, French sociologist and theologian Jacques Ellul wrote, is a "paper pope" that "transforms the freedom of faith into an arrested system that cannot avoid being scholastic in intellectual form."[11] There is an authentic and profound fideism represented by thinkers such as Søren Kierkegaard, Blaise Pascal and Fyodor Dostoevsky, who understood where the proper locus of Christian faith must be and who insisted that believers fully absorb rather than apologetically explain away intellectual challenges to faith so that they might know what a demanding thing faith really is. But the posture of many literalists as the sole possessors of truth and the only believers who take the authority of Scripture seriously begs the question: Is it in fact the authority of Scripture in all of its richness, power, and often enigmatic and untamable diversity that we are being asked to be faithful to? Or rather a rigidified mental system and the unquestionable authority of its self-appointed guardians at any cost?

4

Progressive Versus Degenerating Science

Weighing Incommensurable Paradigms

Before continuing any further we should at least briefly reflect on the question: What precisely *is* the scientific status of "scientific" creationism? "Creation science" is the cognate project of biblical literalism and, according to its most enthusiastic proponents, it provides an intellectually robust, fully credible alternative to evolutionary frameworks. Many Americans believe "creation science" should be taught alongside evolutionary theory even in public high school biology classes as an equally coherent and scientifically compelling way to interpret the empirical data. So what should we make of these pretensions of biblical literalism to rigorous scientific status? Does creationism belong in biology, physics and chemistry classrooms as a serious alternative scientific hypothesis? A brief autobiographical note about my own introduction to evolutionary science might be helpful here.

My first sustained exposure to evolutionary theory, I am somewhat embarrassed to say, only came late in my life, when I took a course in evolutionary biology as a master's student at the Graduate Institute of St. John's College in Annapolis, which despite its name has no religious affiliation and is known for its "great books" curriculum. In keeping with the school's emphasis on the classics, our primary text was Charles Darwin's *Origin of Species*, which we read carefully and discussed closely from cover to cover over an entire semester (supplemented by other important texts in evolutionary theory, from Jean-Baptiste Lamarck to Stephen Jay Gould). In retrospect, I realize I probably often strained the goodwill of my classmates and our very patient tutor (St. John's eschews the title "professor").

I have always enjoyed vigorous debate, and my engagement with evolutionary thinking was at this point in my life still almost entirely defensive and combative, reflecting my uninformed yet deeply held belief that as a Christian it was my duty to be a staunch opponent of whatever Darwin had written. I also assumed that theistic evolutionists had no answers to the kinds of theological objections to evolution I had repeatedly heard raised throughout my life from Christian pulpits and in Christian classrooms, from kindergarten through college. I had made no effort to actually read the writings of any theistic evolutionists. As an undergraduate majoring in English and history I had also never been forced to wrestle with the empirical evidence of evolutionary biology. I simply knew I was right in advance of hearing any reasons why I might be wrong. I needed more time, more experience, and the patience and friendship of others to see that the issues are more complex than I had grown up believing. I still think there are things to be critical of in evolutionary theory, particularly reductive sociobiological ideas in the tradition of Darwin's *Descent of Man*. Yet I can also now see from my own marginal notes in my worn copy of the *Origin of Species* that in my determination to refute Darwin on first reading I often did not even understand what he was saying. I approached the text with an agenda that made me an uncharitable and therefore a *poor* reader.

My primary academic training is in the social rather than the natural sciences, and so I am reliant on trustworthy guides to know where the weight of the evidence points when it comes to things such as radiocarbon dating, plate tectonics, comparative biology, genetics and paleontology. But all scientists are no less reliant than the rest of us on the authority of others when it comes to research in fields distant from their particular areas of expertise. The average nuclear physicist probably knows no more about the courtship rituals of the bonobo apes in the Congo than the average professor of medieval literature. Both must place their trust in the work of experienced field biologists and in established processes of peer review if they hope to learn anything reliable about bonobos. To begin, let us consider some key insights from the philosophy of science over the past 50 years.

The publication in 1962 of Thomas Kuhn's now-classic work, *The Structure of Scientific Revolutions*, itself marked a revolution in the philosophy of science. Five decades later, it remains perhaps the most widely read and cited text in history about the nature of scientific inquiry. Kuhn challenged earlier accounts of the scientific method that reflected notions of linear progress and rational objectivity, including the highly influential ideas of his close contemporary, Karl Popper. Beginning in the 1930s and continuing through the 1960s, Popper championed a definition of "true" science as a field of systematic knowledge acquisition that advances through rigorous empirical testing leading to definitive confirmation or falsification of theories.[1] Kuhn, by contrast, emphasized the rarity of outright falsifications in "normal" science, the persistence of conflicting evidence even in highly successful theories, and the prevalence of scientific "paradigms," that is, research programs rooted in worldviews that might be stubbornly clung to in the face of anomalies. "All historically significant theories have agreed with the facts, but only more or less," Kuhn wrote. Instead of demanding absolute confirmation or refutation of a theory it was therefore best "to ask which of two actual and competing theories fits the facts *better*."[2]

Only when empirical anomalies rise to intolerable levels, Kuhn suggested, do paradigm shifts or scientific revolutions occur. Often, though, rival paradigms exist simultaneously for long periods of time, providing equally compelling accounts of phenomena based upon their own internal logics (not unlike "competing political institutions"[3]). Kuhn's term for incompatible scientific paradigms was *incommensurability*. The Ptolemaic and Copernican models, to cite one example, existed as incommensurable theories alongside each other for approximately two hundred years before the accumulating weight of empirical evidence led to a decisive shift away from the geocentric model of the universe in favor of the heliocentric one. Strictly speaking, though, the geocentric model was never "falsified" in the Popperian sense prior to its rejection by virtually all astronomers. "If a scientific revolution lies in the refutation of a major theory and in its replacement by an unrefuted rival," Greek philosopher of science Imre Lakatos wrote, "the Copernican Revolution took place (at best) in 1838."[4] The Ptolemaic paradigm was perfectly capable of positing new epicycles upon epicycles in an ad hoc way to account for anomalies. Yet its authoritative status was eroded by new astronomical data, which it was unable to account for in a way that was as intellectually satisfying or parsimonious as the Copernican system.

One of the implications of Kuhn's ideas about scientific inquiry being marked by contestation, crises, incommensurability and shifts among competing paradigms is that science must now be seen as part of what sociologists of knowledge have come to call the social construction of reality. Scientific truths are determined as much by historically and culturally inscribed interpretive paradigms (and the array of theoretical and methodological rules that are developed to sustain them) as by their correspondence with the "objective" facts of the universe. *The Structure of Scientific Revolutions* is thus a key text in the rise of antirealist and postmodern philosophies of science (even if this wasn't Kuhn's original intention and although he may well have been reticent about some of the philosophical uses made of his ideas[5]). By bringing to light the paradigmatic nature of all scientific research, Kuhn raised the specter of epistemological relativism in the sciences. His insights led to a seemingly insoluble dilemma: Given the theory-laden nature of all science and the persistence of anomalies in most if not all scientific paradigms, how can we make value judgments between incommensurable but internally consistent research programs? And how can we clearly distinguish between science and pseudoscience, or between science and religion? Does it all in the end come down to a matter of faith—what one *chooses* to believe in—as many creationists claim?

It was in response to the Kuhnian dilemma of judging between incommensurable paradigms and establishing lines of demarcation to tell genuine science from pseudoscience that Lakatos proposed the distinction between what he called *progressive* and *degenerating research programs*. His approach, which in certain ways synthesizes even as it critiques both Kuhn's and Popper's positions, was first published as a long chapter in *Criticism and the Growth of Knowledge* in 1970. Lakatos may be credited with introducing the following key terms to the philosophy of science.

Hard Core

The "hard core" of a scientific theory according to Lakatos is its ultimate explanatory principle, which cannot be changed without modifying one's entire paradigm or research agenda. It is a "negative heuristic" that forbids questioning, redirecting potentially falsifying challenges outward to other theories or doctrines that must "bear the brunt of tests and get adjusted and re-adjusted, or even completely replaced, to defend the thus-hardened core."[6]

Progressive Research Program

A "progressive research program," according to Lakatos, must have a consistently generative or "content-increasing" power. It must (1) yield a steady increase in both theoretical and empirical understanding and (2) offer predictions that are corroborated by unanticipated "novel facts." If more and more such facts emerge that are elegantly and parsimoniously explained by the paradigm, it is a truly rigorous and constructive/progressive research agenda.

Degenerating Research Program

A "degenerating research program," by contrast, is one in which all or almost all auxiliary hypotheses are purely ad hoc in nature, offered simply to protect the hard core in defensive reaction to challenging new data. Degenerating theories are marked by a proliferation of theoretical components, which might be mistaken for theoretical richness or vitality. But on closer examination it is apparent that these ad hoc theories can't keep pace with empirical challenges or new information. They are unable to predict surprising facts or to lead to genuine discoveries. Internal contradictions arise between the ever-multiplying ad hoc theories offered in defense of the core. What may have once appeared as a harmonious orbiting protective shield begins to look more and more like a dangerously cluttered field of space debris.

How do these concepts from the philosophy of science help to shed light on the debate between "scientific" creationism and evolutionary science? First, we should note that Lakatos's framework might very well be used to critique highly reductive forms of materialism, naturalism, evolutionism, scientism or ultra-Darwinism. Christians should have no stake in defending an unqualified Darwinian research paradigm, and many theistic evolutionists are committed no less than young earth creationists to exposing materialism's degenerating character whenever it evolves from a methodological tool into a metaphysical prejudice. (My doctoral dissertation in political theory at the University of Southern California was a critique of the moral and political nihilism of philosophical materialism, with particular attention to the ideas of three exemplary nineteenth-century materialist or naturalist thinkers: Darwin, Karl Marx and Friedrich Nietzsche.) Highlighting problems with philosophical materialism/naturalism and the failure of some popularizers of evolutionary theory to stick to what they do best (namely, sound research science) should not divert us, however, from asking critical questions about the intellectual character of the research program of "creation science."

The question facing creationists is this: Can we honestly embrace young earth or young life creationism as the most rigorous, richly theory-generating scientific research program available, able to account for the current weight of scientific data as well as (if not better than) standard evolutionary approaches, provided only that we understand and accept the creationist's starting assumptions as over and against the materialistic "faith" of the nonbelieving scientist?

The answer to this question is resoundingly clear. Virtually all qualified scientists—including even some dedicated creationists whom I have spoken with—agree that "scientific" creationism has proven woefully unable to keep pace with empirical challenges from a wide array of fields, including geology, paleontology, physics, biochemistry and biology. Creationists have failed to provide convincing models that can coordinate and explain data across diverse fields with the precision and testability that models that allow for "deep time" can. Their paradigms have not led to the discovery of powerfully confirming "novel facts" (the way, for example, that Newton's theory led to the risky prediction of the exact time and place that Halley's comet would return). What creationists *have* done is provide a wide variety of conflicting ad hoc theories to explain facts as they arrive, often resorting to appeals to supernatural interventions when faced with especially difficult scientific puzzles. In fairness, they have proposed some genuine scientific hypotheses, even with modest publication success in noncreationist journals, to explain selected empirical data in ways that are at least *consistent* with their flood geologies and young earth/life models.[7] They have also focused helpful attention on anomalous data that evolutionists at present do not have well-developed explanations for (and have at times attempted to gloss over with dubious "just so" stories).

But as Lakatos made clear, all scientific research programs contain anomalies. Calling attention to challenges to a scientific theory does not invalidate that theory so long as it is still able to explain more data more eloquently than its rivals; and piecemeal findings that might theoretically fit within creationist paradigms do not amount to a refutation of other explanations. The case of Lord Kelvin might serve here as a warning illustration. Based on the best scientific evidence available in the nineteenth century, Kelvin claimed to have mathematically proven that Darwin's theory was false since in the time required for evolution to occur the sun would have burned itself up. Radioactivity and nuclear reactions were beyond Kelvin's comprehension and completely unknown to scientists in his day.[8] Time and again, creationists have similarly committed what we might call the Lord Kelvin Fallacy. They have claimed to discover new evidence that falsifies evolutionary science only to be proven wrong as more data arrives. And time and again, data from multiple independent sources has provided converging support for scientific models based not on young earth or young life models but on deep time and common ancestry. In the refreshingly candid words of creationist and nuclear physicist Ben Clausen, "No comprehensive, short-age model is even available to rival the long-age model. . . . I do not find the evidence for a recent creation compelling. My primary reason for accepting the scriptural account is the part it plays in the Bible's characterization of the Creator."[9]

In short, creationism has been rejected by most scientists not because they are antireligious secular humanists or godless materialists but because creationism is not a robust research program steadily making sense of the world in compelling ways through the power of its falsifiable predictions (which if proven wrong would compel creationists to modify their assumptions about the meaning of Genesis). Scientifically speaking, creationism has all the hallmarks of what Lakatos identified as *degenerating research*. Creationists have made repeated promises that with just a little more time and heroic effort their work will yield resounding triumphs in a wide array of fields.[10] What they are in fact implicitly acknowledging in these perennial statements about being always on the verge of dramatic empirical confirmation is that decades on they are still fighting a scientifically losing (if not decisively lost) battle.

Of course, just as Newton doggedly persisted in developing his theory even when it still had the qualities of a degenerating paradigm, creationists have every right to pursue their research program for as long as their hunches or hopes incline them to do so. Contrary to popular stereotypes, creationists can count among their ranks some very serious scientific thinkers, and one cannot but admire the tenacity and persistence—indeed, the faith—of scholars who hold fast to a failing line of investigation decade after decade—provided only that they do so with intellectual integrity in their methods, with full acknowledgment that the challenges they face are not only scientific but theological and biblical as well, and without any prevarications about how little progress they have actually made in the face of the obstacles. And if someday it turns out that they are right and there *is* strong empirical support for their belief that all life on this planet is in the range of six to ten thousand years old we should cheerfully accept this finding, although our faith should by no means rest upon something as fragile and contingent as this being the case. But one cannot admire or applaud individuals who insist in the face of overwhelming evidence to the contrary that their creationism is currently the most progressive and intellectually rigorous scientific research paradigm available or one deserving of equal time in science classrooms. A progressive scientific paradigm "scientific" creationism is not. Until it becomes so, believing scientists who do not feel compelled to spend their lives trying to reverse the fortunes of degenerating research can and should continue to work with full scientific *and* religious integrity within those frameworks that still make the most sense of the most empirical data.[11]

5

Does Your God Need Stage Props?

On the Theological Necessity of Methodological Atheism

True believers in the literalist paradigm will be unfazed, however, by the degenerating character of "creation science," for the hard core of their research program is ultimately not a scientific but a *theological* one that is immune to empirical falsification of any kind. In the mixed biblical allusions of George McCready Price, "Abraham is commended not because he had things demonstrated for him, but because he 'believed God.' And if we withhold our assent to the truth of God until every possible objection is mastered the plagues will fall upon us unprepared."[1] Creationists have declared that Darwinian theory also cannot be falsified, that it too is based upon "faith." But while they may be correct that we cannot specify a risky test that would positively prove natural selection as the mechanism for macro-level changes in organisms, evolutionary paradigms—whether theistic or naturalistic—differ markedly from "scientific" creationism in that they could very easily be falsified by new empirical data. The discovery of a single rabbit (never mind human or hippo) skeleton in the Precambrian geological strata, evolutionists have long pointed out, would be more than sufficient to completely overturn Darwin's theory.[i] Evolutionary ideas can, in principle, be falsified. If acceptance of the empirical evidence constitutes a leap of faith, it is faith in the intelligibility of the universe open to our investigation.

No imaginable scientific or historical discovery, however, can overturn "scientific" creationism, which declares itself to be a self-validating paradigm or deduction not from scientific evidence but from certain knowledge of the meaning of Genesis. Tellingly, one does not need any particular qualifications as a scientist to establish a name for oneself in fundamentalist circles as an expert in "creation science." It is striking how many of the most vocal champions of "scientific" creationism, from Price up to the present, are not trained research biologists, geologists or paleontologists but rather clerics, church administrators or self-taught science dilettantes who see themselves as courageous knights of faith marching out to do battle against the Philistine Goliaths of peer-reviewed scientific inquiry.

Not a few creationists, though, do have serious credentials as research scientists. We must therefore listen very carefully to what these individuals have to say about the empirical data they have expert knowledge of. We must also—and even more importantly—listen to what they have to say about the kinds of theological beliefs that animate their scientific endeavors. These beliefs, regrettably, often do not seem to be at all theologically sound or well thought out. For example, biologist and biblical literalist Leonard Brand of the Geo-Science Research Institute (a creationist "think-tank" in Loma Linda, California, that employees eight highly qualified research scientists) writes the following with regard to the cosmological difficulties posed by the tenth chapter of the book of Joshua (in which God is said to have caused the sun to stand still at noon "about a whole day" [Josh 10:13] to provide light for the Israelites to annihilate their enemies the Amorites):

[i]No rabbit fossils have ever been found in the Precambrian or Cambrian rocks, although there is at least one glorious hoax online, complete with a photograph of hare bones nestled among the trilobites, allegedly discovered by one "Dr. Wilfred Splenebyrst of the [nonexistent] London School of Ergonomics."

> [H]ow much do we know about the options that an infinite God has at his disposal? And maybe that sun trick wasn't so disruptive after all. . . . A system of giant mirrors could be used to deflect the sun's image, so that from a human perspective the sun did stand still. Then later the mirrors could slowly move the sun back into its normal schedule. Did God do it that way? Of course we have no idea (God is certainly much more creative than us), but this scenario just illustrates how utterly futile it is for finite humans to think we can decide what God can or cannot do.[2]

While this kind of logic may satisfy the yearnings of some Christians for a simultaneously supernaturalistic *and* scientifically plausible literalism (who, after all, can deny that a fully omnipotent God is capable of creating a "system of giant mirrors" to avoid disrupting other aspects of his created universe *if he so desires*?), they raise troubling questions of a much more significant kind—not scientific but rather theological questions.

What sort of God is it whose ways of acting and being in the universe might on occasion require cosmic stage props such as giant mirrors? Is this really the Creator God of Hebrew Scripture and the New Testament witness? Or a deity far closer to the gods of the Greek pantheon with their outsized physical artifices forged in the workshops of Hephaestus on Mount Olympus, exactly mirroring our own merely on a grander scale? Questions like these rarely seem to cross the minds of creationists in their labors to show that Genesis can somehow be read in a way that is indubitably, infallibly and incorrigibly *scientific*. In their preoccupation with establishing the perfect concord of the text with modern scientific criterion of truth, many literalists are in fact not even raising properly theological questions, merely technical ones.[ii]

[ii]The most pressing theological dilemma in the narrative of the destruction of the Amorites and other tribes in the book of Joshua for modern readers, for example, is surely not whether or not we can come up with fantastical ad hoc speculations to somehow sustain its scientific accuracy but the disturbing fact of seemingly divinely authorized genocide—a problem that strict literalism only heightens.

It may be that many creationists have failed to carefully think through the theological implications of their literalism because they have thought of themselves for too long as an embattled minority standing alone for truth though the heavens fall. As a thought experiment, let us therefore imagine that creationists in the near future somehow achieve irrefutable scientific concord between their highly literalistic readings of Genesis and the weight of the empirical evidence. Let us imagine a great paradigm shift occurring in the academy, with creationists coming to be hailed as the most farsighted minds in scientific history. Let us imagine that creationists are able to prove beyond reasonable debate that young earth or young life creationism is the best fit for the evidence across all fields of scientific inquiry and that certain parts of the biological world—leopard eyes, ostrich feathers, etc.—can only be explained as the "irreducibly complex" creations of an Intelligent Designer. What would the theological implications of such a stunning victory for "creation science" be? What would it mean for our faith if incandescent trails of the Creator could, in effect, be incanted by scientists using beakers, flasks or other devices? And what if over time these empirical proofs for an Intelligent Designer became so scientifically clear and testable that any high school student could be walked through a set of lab experiments to empirically demonstrate God's existence, just as easily as they are now taught how to manufacture smoke in a bottle?

This seems to be the fervent desire of some (though certainly not all) creationists as well as intelligent design theorists. Any such deity, however, could only be a greatly reduced divinity—not the self-revealing and self-concealing God of Scripture but a god whom the high priests of science would now have remarkably privileged access to and unprecedented powers to dispense. "Scientific" creationism—the kind of creationism that is unwilling to live with tensions between the language of Scripture and what the weight of the scientific evidence tells us about natural history—at its heart may be the equivalent of an unholy longing for *scientific transubstantiation,* a desire to see the wine of communion turning to the plasma of Christ beneath the world's most powerful microscopes.

These perils were grasped by two of the most devout theistic scientists in history. The move to exclude "natural theology" from science was first championed not by Darwin but by Isaac Newton and Robert Boyle, who saw that the metaphysical mixing of modern empirical methods with religious teleology resulted not only in bad science but also in a corruption of true faith. God's transcendence *theologically* requires a radical distinction between God as Creator and the operations of the universe through secondary causes that can be empirically observed and tested through inductive and deductive methods. Methodological atheism was necessary, Newton and Boyle maintained, not to protect science from religion but to protect theology from diminishment, trivialization and manipulation by scientists.[3]

There are nevertheless good reasons to think of theology—in a very different sense than creationists have conceived—as a "scientific" endeavor that pursues systematic knowledge acquisition and that includes (in Kuhnian terms) incommensurable paradigms and (in Lakatosian terms) both progressive and degenerating research programs. A progressive theological research agenda will readily be distinguished from a degenerating one by its ever-expanding compass of fresh insights and its ability to make surprising sense not only of long-standing difficulties but also of unforeseen challenges and new data as they arise. It will lead to the recognition or discovery of new theological insights and provide vital inspiration and resources for their elaboration. Its primary mode is not one of defensive or fantastical ad hoc apologetics but rather of fearless exploration in openness to new knowledge from a wide array of sources. In this light, it is hard not to conclude that "scientific" creationism and strict literalism on Genesis represent not only a degenerating scientific paradigm but a degenerating theological paradigm as well.

Perhaps the most widely deployed auxiliary theory in the protective belt that encircles strict literalism on Genesis is the claim that only young earth or young life creationists take seriously the authority of Scripture. Yet if the lives and witnesses of actual believers matter at all to our thinking, this claim is demonstrably false. Numerous believers with unimpeachable credentials as scientists, as theologians and as biblical scholars, hold very high views of the Bible's authority while embracing nonliteralistic readings of Genesis. What they have challenged is not the inspiration or authority of Scripture but the appropriateness of rigid hermeneutical approaches to the Bible that treat the creation narratives as a scientific-historical record. There is always, of course, the possibility that these individuals are mistaken in their readings. No one's ideas should be treated as being beyond thoughtful criticism. But the fundamentalist insistence that these committed Christians can only be one of three things—mentally feeble, morally suspect or spiritually deficient—is perhaps the most depressing illustration of how degenerating the linear equation of literalism on Genesis with belief in biblical authority has become in much creationist discourse. This hypothesis can only be sustained if we cloister ourselves behind very high walls lest we encounter the actual lives and thinking of others. (A fundamentalist can never be too careful what she reads or whom she befriends!)

It is strict literalists themselves, we have also seen, who have most clearly subjected the theological authority of Scripture to the authority of modern scientific rationalism with their insistence that Genesis *be* "scientific" in order to be divinely inspired. This has led to unconvincing ad hoc theorizing about evidence even within the biblical text itself, such as attempts to force complete chronological agreement between Genesis 1 and 2. Believers ought not to be so in awe of modern science as to assume that Genesis is only authoritative so long as it is scientific or historical. Or what of the creationist assertion that the core of *their* theology, in contrast to others, is belief in God as Creator? Certainly no one doubts the depth of literalist concern for the centrality of belief in God as Creator to Christian thought. But since nonliteralists also believe in God as Creator, this again cannot be the actual core of literalist theologizing, unless we add a key modifier that turns the entire proposition into a pure tautology: the paradigmatic hard core of literalism on Genesis is belief in *the literalist God* as Creator. Or what about the literalist assertion that "creation science" perfectly coincides with the original meaning of Genesis intended by its author? There are, again, a great many biblical scholars who say otherwise. At the least literalists must acknowledge that it is a live question. Or how about the literalist claim that if we deny "scientific" creationism on scientific grounds we must also for the sake of full consistency deny the historical resurrection of Christ? This is a complete non sequitur, logically equivalent to saying there can only be fast miracles, not slow ones. "The medieval wizard may have flown through the air from the top of a tower," G. K. Chesterton wryly observed, "but to see an old gentleman walking through the air in a leisurely and lounging manner, would still seem to call for some explanation."[4] Theistic evolution is, we might say, leisurely creationism.

There is one (and to my mind only one) theologically serious reason why one might hold fast to a strictly literalistic or "scientific" reading of Genesis: the concern that evolutionary paradigms raise the theodicy problem of animal suffering to an intolerable level. When not consumed by thoroughly rationalistic and scientistic habits of mind, creationists pose a genuine theological objection to the idea of evolutionary creation: How could a God who created through a process that involved death and suffering *from the beginning* be described as both fully just and fully loving? What do evolutionary paradigms say about God's character? To state the challenge in the form of what Lakatos called a "positive heuristic," the literalist research program rests in part upon the following (at least initially) plausible claim: *Only a strictly literal reading of Genesis that attributes all mortality and all predation in nature to Adam's fall can explain the origins of animal suffering in a way that maintains God's justice, God's omnipotence and God's love by refusing to allow that the Creator is the author of the evils of pain and death.*

I will take up this challenge (and some of the ideas about the meaning of Christ's life, death and resurrection that accompany it) in part two of this book. In the meantime, we may summarize the literalist position as follows. The reason literalists read the creation narratives and other parts of Scripture the way they do is because they are already committed to a very specific philosophical and theological research program, namely, to a kind of foundationalism that owes its lineage to the ideas of Descartes and other Enlightenment thinkers as much if not more than to the ideas of Scripture. The burning heart of modern creationism is not a doctrine but a *method*. Doctrines will be creatively reinterpreted or even rewritten without hesitation by literalists in order to sustain this methodological project. What must be protected from change at all cost is the paradigm of philosophical foundationalism-cum-literalism itself, which grounds the literalist's sense of certainty and security in an uncertain age. To change *this* would mark the collapse of the literalist's research program and require a significant paradigm shift—and there is no greater fear among creationists than the fear of paradigm lost.

6

The Enclave Mentality

Identity Foreclosure and the Fundamentalist Mind

When believers sever themselves from the long tradition of both Protestant and Catholic as well as Jewish nondogmatic and nonliteralistic approaches to questions of origins, the result is *fundamentalism*. I have already used the word several times without defining it, and no settled definition exists. *Fundamentalism* is a contested term that has often been used for purely polemical purposes. It clearly has a strong negative connotation in the minds of most believers and nonbelievers alike. Others, however, embrace the label as a badge of honor and a mark of defiance against what they see as a corrupted and corrupting theological liberalism and faith that is too open to scientific reason. (Years after the fact George McCready Price would boast that his 1902 work *Outlines of Modern Christianity and Modern Science* was "the first Fundamentalist book."[1]) Historically, fundamentalism as a self-identified movement arose in the United States in the early part of the twentieth century as a reaction to the challenges posed by modern science and especially by evolutionary ideas. This raises the question of whether it is legitimate to apply the word to Muslims, Jews and others, or to Christians from earlier time periods.

A word of caution is also in order: not all fundamentalists are literalists on the creation stories in Genesis, and not all literalists are fundamentalists; Charles Hodge, the conservative nineteenth-century theologian who played perhaps the most influential role in the rise of American fundamentalism, interpreted the days of Genesis 1 as indefinite lengths of time. Nevertheless, biblical literalism and fundamentalism have in practice been largely synonymous in the United States for at least the past five decades. I use the term here in broad sociological and psychological perspective to describe any groups or individuals who hold to notions of scriptural inerrancy or infallibility (whatever their Scripture may be) across all fields of human knowledge, including history and science, in reaction against the moral and epistemological challenges of modernity. The Bible is free of error and self-interpreting, fundamentalist Christians declare, so that we must reject readings that are influenced by new scientific or historical discoveries or that seek to contextualize its narratives in terms of the beliefs and literatures of the time period in which they were written. Instead, all empirical and historical evidence must be subordinated to what the Bible is alleged to say with absolute clarity and timeless authority about itself.

"Creation science," according to literalists, is superior to any merely rational, empirical and inductive science, amounting instead to a species of deductive reasoning to uphold an unfalsifiable, foregone conclusion. To outsiders, the results may appear (at best) as a mode of critique without constructive alternatives and (at worst) as an intellectually dishonest form of data mining. But for the true believer (as Eric Hoffer referred to the followers of ideologically driven movements), intellectual honesty does not mean following the empirical and inferential trails wherever they may lead. It means holding fast to the inerrant words of the sacred text at all costs. Where others *interpret* the Bible according to their fallible human reasoning, literalists (in their own self-understandings) faithfully *accept* the Word of God and then courageously seek out the evidence to defend it. This means that those who disagree with fundamentalist interpretations of the Bible, no matter how seriously they take the Bible's authority, are often charged with being faithless individuals in rebellion against divine truth itself. What we witness in fundamentalist communities and individuals is the move from total confidence in one's ability to understand the "literal" truth of

Scripture to relentless suspicion and even demonization of those who read the sacred text in other ways.

In my own tradition this has included increasingly clamorous calls by some believers for the establishment of a de facto Magisterium (though they would never use the word) with the power to pry into the consciences of individuals and, in the case of scientists and theologians under church hire, to punish any openly expressed nonliteralist beliefs about Genesis as an intolerable mark of (variously) weak faith, denominational disloyalty, intellectual deficiency and/or moral turpitude. What these individuals plainly desire is not a doctrine of creation but a *dogma* of creation*ism*. Oliver O'Donovan's lament over the state of moral discourse in the Anglican Communion (on both the "left" and the "right") surrounding questions of gay marriage might just as well be applied today to the Adventist Church's handling of questions of evolutionary biology:

> Stepping back, untangling the skein, reconciling conflicting views, toning down exaggerated positions, forging coalitions, squaring circles, finding commonsense ways through: the whole stock in trade of a tradition once defined by opposition to enthusiasm of every kind, seems to have been mysteriously wiped off the software. In its place are radical postures, strident denunciations and moralistic confessionalism.[2]

The *fideles* must be exceedingly careful about even entering into dialogue with those who read the Bible and see the scientific evidence differently than they do, since to entertain questions about "the fundamentals" is to sup with individuals who may quite literally have lost their souls. Any dialogue that occurs will, for the true believer, be with a view to converting the recalcitrant other to the truth that is already known, or else impressing upon any wavering onlookers the dangers of the nonfundamentalist's ways. Authentic dialogue, in the sense of an open-ended, mutually risky and nondogmatic search for greater understanding, cannot be permitted, since this would imply that there are legitimate questions to be asked about the very thing that for the fundamentalist cannot be questioned: the absolute convergence of this method of reading the Bible with God's perfect and epistemologically exhaustive revelation. Fundamentalists seemingly lack all critical self-reflexivity. They are never in their self-understandings engaged in fallible interpretations of Scripture on the basis of complex and contestable sets of culturally and historically inscribed assumptions and philosophical commitments. They are always and only declaring what the text self-evidently means. Their reply to challenging alternative readings thus repeatedly takes the form of arguments from personal incredulity: "How can you possibly disagree with me? The fact that my reading is the one true reading is *obvious*."

The fundamentalist can barely fathom, let alone practice, the irenic stance of a towering and generous theological mind such as Karl Barth toward what he himself regarded as a "heresy" within the body of believers. In reply to a letter from a devout pastor demanding that liberal scholar Rudolf Bultmann be removed from his position as chair of New Testament at Marburg University for his denial of the physical resurrection of Christ, Barth wrote, "A contesting of heresy which misses the essential point . . . has always been more dangerous to the church than the heresy in question."[3] The "existence of a 'heretic' like Bultmann, who is so superior to most of his accusers in knowledge, seriousness, and depth," he continued, "might be indirectly salutary to the church as 'a pike is in a pond of carp.'" By contrast, "the rise of a clerical group that hands out censures or doctrinal judgments with neither true vision nor reflection, no matter how orthodox it might aim to be, can only bring about destruction." (If nothing else, fundamentalism is, we might say, the longing to live in a pond of one's fellow carp without any pike.)

The spirit of censure and craving for communal purification is ultimately what makes fundamentalist readings of Genesis (as well as other parts of Scripture) *fundamentalist* as opposed to merely literal. The fundamentalist does not simply have a view of what Scripture means, arrived at through many years of prayer, meditation, reflection and careful study, humbly held and generously shared with others. Rather, she denies the very possibility that there can be other faithful ways of reading the sacred text using different hermeneutical lenses. He insists that others are not simply mistaken but that they must be morally dishonest, intellectually deficient or spiritually bankrupt for refusing to accept the self-evident truth. She strives not simply to persuade others with winsome reasoning in a shared quest for new light but to silence those she believes are corrupting the community (and especially the young) with ideas not based upon God's infallible Word but corrupt human science and reasoning. Fundamentalism, as a militant, mobilized reaction to the challenges of modernity, therefore assumes the character of ideology, agitprop and power play. Its rhetorical mode is best captured in the simplifying tract for the masses, the televangelistic blitzkrieg campaign, and the revivalist "crusade" centered on the charismatic authority of a powerful speaker who is able to sway his audience to action as he confidently strides back and forth across the podium reeling off texts, cajoling his listeners and smiting his Bible (King James Version preferred).[i]

Fundamentalism, then, is not simply a way of reading texts. It is a plan for political action. And fundamentalist political action in secular as well as ecclesial realms has often led to violence, whether in the form of the righteous crusade against "heathen" outsiders or the scapegoating of "heretical" insiders. Once it becomes clear to the fundamentalist that he cannot win the day by citing verses alone since others stubbornly read the same verses differently than he does, he will move to create a centralized political power or ecclesial body with the authority to suppress rival interpretations, to monitor for unacceptable thoughts, to denounce infidels and to vigilantly police the boundaries of the community. Yet even as the fundamentalist sows great destruction and inflicts real violence on the Other (whether psychological or physical, ranging from ostracism and excommunication to pogroms and holy wars), he invariably thinks of himself as a *victim* of the Other's aggression. "In their own eyes, the fundamentalists are reasonable

[i]The speaker is almost always a *he*, with most fundamentalist communities viewing women in the pulpit with high levels of suspicion if not outright disapproval.

people," Emmanual Sivan writes. "It is just that committed to revelation, they are pitted in a fight against the outside. . . . Rational yet embattled, it is a self-perception no doubt as sincere as it is deep."[4]

The mere fact that others disagree with the fundamentalist's interpretations and openly offer other ways of thinking about the text is felt by the true believer as a direct existential threat to themselves and to the entire community—a sinister danger that must be exposed and cleansed. This is especially true if the Other claims to also be a believer and faithful member of the tribe. The enemy within is deemed far more dangerous than the enemy without and is held in special contempt by fundamentalists as one marked not simply by error but by "infidelity" (analogous to what is known in Marxian theory as "false consciousness" and in Islamist ideology as *Jahiliyya*). As a form of foundationalist philosophical reasoning, fundamentalism declares that failure to hold fast to the "correct" interpretation of any one of the fundamental beliefs must necessarily unravel all of the others, spreading rings of contaminating influence throughout the community and finally toppling the entire faith. As a totalizing political narrative (and the "fundamentals" tend to increase over time, encompassing ever more of the individual's life), fundamentalism declares that the dissent of even one member pollutes the entire body.

But fundamentalist readings of Scripture are of course precisely that: *readings* that may be challenged not only on scientific, historical and philosophical grounds but on theological and biblical grounds as well. God may not be fallible, but fundamentalist hermeneutics and exclusionary logic certainly are. Fundamentalism is an idolatrous form of human reasoning, both from and about texts, not because it takes Scripture literally but because it totalizes its literal readings while denying the very possibility that it might be wrong. Fundamentalisms arise, Edward Farley writes, "when religious leaders so work to protect the perennial (authoritative) mediations of religion from the modern that the mediations become themselves the contents of religious faith."[5] Another way of stating this would be to say that fundamentalism presents itself as the height of faith when it is, in the deepest sense, an expression of faithlessness.

We can now see that fundamentalism, in its often uncompromising literalistic stance toward Genesis, has as much to do with the peculiar psychological makeup of individuals as it does with abstract theological reasoning. We can perhaps better understand the fundamentalist mind in terms of concepts developed by psychoanalyst Erik Erikson, who famously suggested that psychologically healthy individuals must successfully navigate at least eight developmental stages between birth and death.[6] One of the most critical periods in this process occurs between adolescence and young adulthood. In order to gain a clear sense of personal identity, Erikson suggested, we cannot simply follow the prescribed paths laid out by our elders but must explore other possibilities, challenging perspectives and alternative values. The final result of this period of searching may well be a decision to retain much of what we have received from an early age. But it is essential, Erikson believed, that this choice be based upon one's weighing of multiple perspectives and wrestling with life's complexities for oneself. Some individuals, however, never pass through this critical stage of identity formation (or else do so only very partially). Erikson's term for this was *identity foreclosure* or *premature integrity*.

Identity foreclosure can happen for a number of reasons, including an existential crisis or trauma such as an experience of social marginalization or the death of a loved one. Typically, though, individuals with foreclosed identities are persons whose sense of self-esteem has been highly dependent from an early age on the approval of strong authority figures, especially their parents and most often their fathers. They are men and women who experience unusual pressures in childhood to conform to the values and expectations of others. Having been denied the "moratorium" or leeway in adolescence to delay adult commitments, they prematurely embrace a set of values that was forged for them.

Individuals who have undergone identity foreclosure might be highly successful in many ways. But their personalities are also marked by destructive psychological traits. They are less self-reflective than others. They are often mentally rigid, tending to see the world in terms of simplifying narratives that are beyond question. They are incapable of incorporating new values or perspectives into their worldview. They have difficulty cultivating warm and intimate relations even among some of their closest friends and loved ones. They have little patience with ambiguity and little intellectual curiosity in unfamiliar ways of thinking. They seek refuge in overarching meaning structures that are uncompromising and total. They are often deeply concerned with maintaining authority structures and upholding traditional religious values. People with foreclosed identities are thus naturally drawn toward fundamentalist communities—and in an insidious feedback loop, fundamentalist communities produce people with identity foreclosure. Erikson's categories might apply, then, not only to individuals but to social groups as well: fundamentalist religious communities are communities of identity foreclosure in which internal tensions and contradictions are resolved at the steep cost of *institutionalized* premature integrity.

Many readers will by now have already detected the great internal contradiction in the fundamentalists' claims that they are defending the authority of Scripture alone; for what is required in actual practice to maintain their interpretations of *sola scriptura* are an office and a tradition (and often, for safe measure, a charismatic leader or prophet as well) possessed of the extrabiblical authority to prescribe both orthodoxy and orthopraxy for the community. "It is the destiny of the *sola scriptura*," theologian John Milbank writes, "to be so deconstructed as to come to mean that we must believe the Scriptures because they are politically authorized."[7]

There is one way that Christians might possibly prove Milbank wrong: they might come to see the meaning of *sola scriptura* not as they traditionally have in terms of the supreme trumping or silencing power of the text over all potentially rival sources of knowledge, but rather in terms of the noncoercive, nonviolent authority of a witness that continually evokes and invites open-ended conversations about its meanings. The unique or sole authority of Scripture would in this case lie not in the Bible's ability to provide once-and-for-all answers to all our most pressing questions, nor in its power to tidily dispatch with challenging new evidence from history, science and human experience with a fideistic "Thus saith the Lord." It would instead lie in the way that reading Scripture in communion with others who are also committed to making its narrative central to their lives teaches us how to ask and re-ask the right kinds of questions in the midst of new realities with the right kind of responsiveness to the Other with whom we may have unresolved—possibly unresolvable—disagreements.

One can only be true to the principle of *sola scriptura*, I am suggesting, if one is also committed to the principle that the biblical narratives—precisely because they *are* narratives as opposed to rationalistic syllogisms—draw us into a continuous dialogue that must always strive for the greatest possible inclusion of voices among those who continue to desire to be part of the conversation. Scripture's truth is often disclosed more in the character and quality of this ongoing exchange *about* Scripture than in the crowning of "winners" and "losers" (or worse yet, the branding of friends and enemies) in zero-sum debates about final, exhaustive meanings.

And the way that we talk to one another about Scripture is one of the ways that we ourselves become part of Scripture's story, whether for good or for ill, revealing what kind of authority Scripture is actually playing in our lives. "When people in the church talk about authority they are very often talking about *controlling* people or situations," N. T. Wright observes. However, he continues, "there is no biblical doctrine of the authority of the Bible. For the most part the Bible itself is much more concerned with doing a whole range of other things rather than talking about itself."[8] For example, the New Testament is centrally concerned with the story of how the guardians of right religion colluded with the political authorities of the empire to kill a blameless man who threatened their deeply ingrained assumptions about who God was, about the meanings of their holy texts and about their own special standing as the sole authorized dispensers of truth. So how should *this* story—the story of Christ and the Pharisees—function as authoritative for us today? When we disagree with others about questions of origins, where do we find ourselves within its narrative arc?

Unfortunately, the fundamentalist understanding of biblical authority, by definition, leaves no room for interpretive differences where questions about passages like Genesis 1 and the relationship between Scripture and scientific evidence arise. The kind of authority many Christians want for the Bible in general and for the creation narratives in particular—the authority of an inerrant Answer Book to stave off the threat of modern materialism—reflects not a higher but rather a *lower* view of Scripture's authority. "The problem with all such solutions as to how to use the Bible is that they belittle the Bible and exalt something else," Wright points out. "Basically they imply . . . that God has, after all, given us the wrong sort of book and it is our job to turn it into the right sort of book by engaging in these hermeneutical moves, translation procedures or whatever."[9] Wright describes these conservative evangelical and fundamentalist approaches to the text, sincere though they may be, as in the final analysis *sub-Christian*. And sub-Christian visions of the Bible—even those that hold aloft the banner of *sola scriptura*—can only lead in the end to a subordination of biblical authority to the realm of the political, just as Milbank charged.

7

The Gnostic Syndrome

When Literalism Becomes a Heresy

The perils posed by "scientific" creationism and dogmatically literalistic readings of Genesis to a vibrant Christian faith regrettably do not end here. The hazards are not only hermeneutical (the danger of doing violence to the "plain" meaning of the text in the name of defending it); philosophical (the danger of overcommitment to an outmoded, thoroughly modernist ontology and foundationalist epistemology); sociological (the danger of answering challenges to one's worldview not with intellectual honesty and careful foresight but with simplistic slogans, fideistic dismissals of difficult evidence, and the exclusionary praxis of fundamentalism); or psychological (the danger of destroying the very possibility of a rigorous discipleship of the mind by shutting down pathways of investigation before they have been fully explored, leading to identity foreclosure and premature integrity). There is another very real danger in fundamentalist forms of creationism, and that is the *spiritual* danger fundamentalism poses to fundamentalists themselves. Christ warned his disciples in the strongest possible terms of the ironic reversal that occurs when individuals set themselves up as spiritual judges over others (Mt 7:1-5). In their zeal to define others out of the life of Christian faith (or out of their particular enclaves of true belief), fundamentalist creationists themselves can quickly come to exhibit all the marks of a very ancient heresy.

In his Memorial Lecture at King's College in London in 1944 titled "The Inner Ring," C. S. Lewis provided an astute analysis of one of the most powerful impulses in human life: the longing to be welcomed into some special, inner circle. All societies, Lewis declared, are structured as concentric rings and rings within rings. These groupings include formal and informal, written and unwritten, rules of admission. For those who find themselves situated in the outermost rings of any social group or culture, there is inevitably a strong desire to somehow ascend to the higher or innermost circles of power, prestige, fortune, comfort and friendship. For some, however, the longing to be part of an "inner ring" moves in the opposite direction from what one might predict. What these people most yearn for, Lewis wrote, is "some artistic or communistic côterie . . . the sacred little attic or studio, the heads bent together, the fog of tobacco smoke, and the delicious knowledge that we—we four or five all huddled beside this stove—are the people who *know*."[1]

Although he did not directly refer to gnosticism in his lecture, Lewis's linking of the pervasive human experience of social and existential alienation with both political ideology (the seductive thrill of the "communistic côterie") and the longing for secret knowledge (the desire to be part of an elite group "who *know*") sheds important light on the gnostic temptation in all its diverse historical guises. The masses may envy the inner circles of kings, but others dream of being part of the ring of true prophets—of tearing down the false gods and raging against thrones and false priests. Among these radicals a still smaller circle is irresistibly drawn to that flame we might refer to as *the Gnostic Syndrome*. I am not the originator of this phrase but have borrowed it from the work of political philosopher Luciano Pellicani, whose book *Revolutionary Apocalypse* provides an insightful analysis of the sociology, psychology and history of Marxism as a form of secular eschatology. On first reading his description of the utopian political radicals of the nineteenth and early twentieth centuries I was struck by how similar they sounded in their pathos to many creationists I have known.

The idea that at least some forms of creationism bear a striking family resemblance to gnosticism may at first sound incredible to readers familiar with the teachings of the gnostic movements of the first and second centuries. The original gnostics disparaged all material existence, including the biblical doctrine of creation, placing them in certain ways at the farthest possible remove from contemporary creationists.[2] According to the Samarian Simon Magus, who was a contemporary of the apostles, the world is the unauthorized evil creation of a rebellious angel or demiurge. Simon, together with his consort Helena—a prostitute he had discovered in a brothel in Tyre and whom he proclaimed to be the female hypostasis or incarnation of the divine Spirit—had been sent as Messiah to release the divine spark within humans so that they might be reunited with the Unknown God, the unfathomable "One, root of the All."[3] The gospel according to Marcion, who was excommunicated from the church of Rome in the second century, bore a closer resemblance to orthodox Christianity but was similarly anti-Jewish at its heart. The earth, he taught, was the creation of an oppressive lower-order world-god. Christ was sent by a higher divinity in order to deliver humans from the realm of mere tellurian, temporal existence. The task of believers was not to sanctify life in this world through acts of justice and mercy as in the Jewish faith, but rather to reduce contact with the world as much as possible through ascetic practices and social withdrawal in order to achieve spiritual perfection. In the gnostic speculations of the second-century Egyptian Valentinus, the demiurge responsible for creating the world appears in an only slightly more positive light. The God of Jewish Scripture, according to the Valentinians, foolishly and conceitedly (though perhaps not evilly) *thinks* he is the highest or only God. However, there is another above him: Sophia, the Mother goddess of Wisdom. Above her there is another still, Bythos, the ineffable Absolute from which the Pleroma (the totality of divine beings) emanate. Again, the goal of life for those few who are able to discern the truth is to escape corrupt material existence through the pursuit of secret knowledge or *gnosis*.

Remote as these specific doctrinal beliefs may be from the creationisms of today's fundamentalist Christianities, however, gnosticism is best understood not as a specific set of doctrinal formulations but rather as a general orientation toward the problems of suffering and alienation. As such, it remains an enduring temptation in Christian and even secular thought. Hans Jonas located the gnostic spirit in the modern world in the philosophies of Friedrich Nietzsche and Martin Heidegger, with their proclamations of cosmic abandonment and attempts to find a new way of being human—accessible only to a tiny minority—to escape the maw of nihilism that opened with the alleged death of God.[4] Eric Voegelin detected a return of the gnostic religion in the totalitarian political ideologies of the twentieth century.[5] Pellicani likewise emphasizes the need to focus on the underlying and perennial human experiences and longings expressed in the strange visions of ancient gnostic soteriology and eschatology. "Before being a doctrine," he writes, "Gnosticism is an existential disposition of the soul that covers every aspect of life: conduct, destiny, a person's very being."[6] The Gnostic Syndrome is thus a complex of feelings and convictions about reality that might take both secular and religious forms but that prototypically includes the following elements (which I have taken some liberty in distilling and elaborating from Pellicani as well as Voegelin).

Anxiety

Gnostics are filled with a painful fear of cosmic abandonment. They know the dark night of the soul, the depths of human depravity and injustice, the anguish of great disappointment in the seeming absence of God. They feel compelled to continually ask the most pressing metaphysical questions: Where do we come from? Why are we here? What is the source of human suffering and evil? Why am I unhappy? Will there ever be a total answer to these problems? What is morally required of me to help usher in the final harmony? Is all this but a cunningly devised fable? How long must we wait?

Alienation and Suspicion

Gnostics are filled with nausea when they survey the world in its presently existing state. To a much greater extent than others they feel *in* the world but not *of* it. Society is radically corrupt. Its political, religious and cultural institutions are cold, unfeeling monstrosities that cannot answer the gnostics' innermost spiritual and psychological needs. They are convinced that these institutions are in fact dominated by evil forces and so must be viewed with constant wariness, suspicion and distrust.

NOSTALGIA

Gnostics conclude that the reason they are unhappy is that they have somehow been thrust into an abnormal situation. This world is not the true reality but a perverse inversion of the natural order. It was only through a catastrophic cosmic accident that they became trapped victims in this veil of tears. Hence, Pellicani writes, the gnostic "is dominated by a desperate *nostalgia* for a *totally different* world, which he has never seen, but from which he feels unjustly exiled."[7]

MILLENARIANISM

All history is divided by gnostics into three stages or eons: (1) a lost age of perfection; (2) the present age of alienation in which we now find ourselves confined as though in a prison; and (3) the soon-coming age of paradise restored. We are living in the final days of the second age, the age of alienation, which must be transcended, annihilated and overcome through a historical process involving intense, conflictive struggle.

DUALISM

The world is conceived by gnostics as a great battlefield in which the "children of darkness" face off against the "children of light." Physical matter ("the flesh" or other material factors) interferes constantly with our ability to discern the true nature of existence and achieve perfection.

PERMANENT REVIVALISM

The gnostic's suspicion of the world around him is matched by a boundless optimism in his own capacity, together with a small group of the elect, to take action in a way that will catalyze the transformation of the fallen world into the final harmony. It is fully within the capacity of human beings to overcome the effects of the fall, both internally and externally. "Life is therefore a state of *permanent waiting* for radical renewal."[8] On the path of perpetual reformation or perpetual revolution, all that remains is for the vanguards of the millennium to exert one final herculean effort to usher in paradise (and inevitably, when this effort fails, another . . . and another . . . and another).

Elitism

In gnostic thinking humanity is divided into three classes: (1) the gnostics or *pneumatics* who are the chosen few in possession of the hidden knowledge or true light necessary to achieve perfection/salvation; (2) the blind masses who might still obtain salvation but only if properly guided by the gnostic elites; and (3) the corrupted ones or children of darkness who do not possess the truth and who exert a sinister influence on the masses. This final group is marked for destruction in the coming conflict. As a class, the leaders of gnostic millenarianism are typically middling intellectuals—educated persons frustrated by a sense of exclusion from the inner circles or higher echelons of learning (which they therefore hold in contempt) and confident that they possess the mental talents to lead great historical movements. They bear a striking resemblance to the group Norman Cohn described as the *prophetae* or "freelance preachers" who led repeated fanatical apocalyptic revivals during Europe's Middle Ages.[9]

Salvation by Knowledge

The gnostic solution to human unhappiness is *gnosis*: complete knowledge—descriptive and normative—of the origins of the fall, the scandal of evil and the pathway to redemption. The gnostic possesses a comprehensive understanding of who the children of darkness are, of why the world of matter cannot be trusted, and of when and how salvation will occur. This knowledge, hidden from others, is itself the means of the gnostic's salvation. Knowledge of the course that history must run from beginning to end is what transforms the gnostic from a person of "mere" faith into a revolutionary filled with the missionary zeal of fanatical certainty. Whether formally acknowledged or not, one is saved in gnostic soteriology not by *pistis* (faith) but by *gnosis* (knowledge) that serves as a "liberating science" or "diagnosis-therapy" of the human condition and counter-explanation of material realities.

Surrealism

The "liberating science" of gnosticism is not the disciplined practice of empirical inquiry in openness to the world of material facts as they often stubbornly confront us, holding us accountable to reality in at times discomforting ways. It is instead a revolutionary "science" that frees the true believer from the "false consciousness" associated with ordinary science, fallen human senses and rationality. Only those armed with the special hidden knowledge are able to correctly "read" material reality. Gnostic epistemology thus assumes the character of an "ideological surreality," a totalizing discourse that cannot be refuted by any encounter with reality whatsoever.[10] The point is not simply that gnostic "science" is unfalsifiable. It is that gnosticism is a relentlessly self-referential framework that is ultimately not bound by the rules of ordinary science at all. "In Gnosticism the nonrecognition of reality is a matter of principle," writes Voegelin.[11] In the case of orthodox Marxism, Pellicani observes, "theorems can never be refuted by reality"; all rational or empirical objections to Marx's "scientific materialism" are dismissed as meaningless "pseudoscience" by the "guard dogs of orthodoxy."[12]

Authoritarianism and Absolutism

Those who claim to be part of the charismatic community but who do not exhibit unconditional loyalty and cheerful compliance to the values and beliefs of the group as promulgated by its central authorities or charismatic founding prophet must be excommunicated or eliminated. The gnostic church or party "cannot tolerate the existence of traditions, institutions, resources, interests that do not fall within its normative jurisdiction."[13] There is no room for tolerant pluralism where challenges to gnostic "science" arise, nor can doctrines be interpreted by members for themselves—they can only be correctly interpreted by the gnostic intelligentsia, whose absolute authority is essential for the "pedagogical protection" of the masses.[14]

(Elaine Pagels has suggested that the original gnostics of the first and second centuries represented a more tolerant, feminist form of Christianity whose texts were excluded from the biblical canon by a conspiracy of oppressive male priestly elites. According to Hans Jonas in his classic study *The Gnostic Religion*, however, the establishment of the Christian canon was largely an answer to Marcion's attempts to impose his own gnostic canon upon the church that would have expunged the Hebrew Bible from Christianity. It was the gnostics themselves who appear to have exhibited the most censorious tendencies with their extreme contempt for Jewish beliefs about the goodness of creation and material existence. The fixing of the Christian canon thus had as much to do with fighting to keep the Hebrew Scriptures *in* the Christian Bible as with trying to keep other books out.[15])

Isolationism

The goal of gnostic education or "pedagogical protection" is not to open new windows outward to the world so much as it is to construct high walls to keep the realities of the outside world from breaking in. The only way to guarantee the doctrinal purity of the movement is through intensive indoctrination, vigilant monitoring of thoughts, and where possible the removal of the core membership from their surrounding environments. The revolutionary fervor of the gnostic party can only be sustained by keeping members in a state of mental, spiritual and even physical/geographical isolation. The only voices they should hear—the only books they should read, the only speakers they should listen to—are the voices of dedicated fellow travelers (or those approved as "safe" by the leaders of the party).

LACERATION

The periodization of all of history into the tripartite division of paradise/paradise lost/paradise restored is intensified and retold in gnostic soteriology to give meaning to the historical frustrations and setbacks of the children of light themselves. There is not merely one fall but falls within falls. The restoration of the golden age begins with the perfection of a small group who has grasped the hidden truths of history; but as time goes on and injustice and unhappiness persist, the gnostic critique of the world turns inward into self-critique and self-laceration. The gnostic becomes convinced that the children of light have themselves somehow been corrupted by "deviationists" in their midst. These untrue ones are delaying the forward motion of history. They must be purged or "shaken out." The alleged perfection of the pioneers must be recaptured by the gnostic community before perfect harmony can be brought to the world as a whole.

Some elements of the Gnostic Syndrome, we must note, are very close if not identical to the most orthodox Christian beliefs (as will be clear in my discussion in part two of C. S. Lewis's "cosmic conflict" approach to the theodicy dilemma of animal suffering). But whenever the above elements are found in *conjunction*, we are confronted with a belief structure, psychology and social practice that is gnostic at its core. What, then, does the Gnostic Syndrome have to tell us about "scientific" creationism, not merely as an abstract doctrinal system but also as sociological phenomenon and existential disposition of the heart and mind?

Conor Cunningham locates the root of all gnosticisms in dualistic thinking and charges creationists with invariably succumbing to an essentially gnostic opposition of the material and the spiritual, the natural and the supernatural.[16] This strikes me as too broad an indictment that does not correspond to the lives and thinking of many creationists I know. At the same time, I think Cunningham is on to something. Not a few creationists I have met strike me as being deeply entangled in the Gnostic Syndrome. Few enduring religious traditions will be as extreme as the ideal-type Gnostic Syndrome I have outlined above. Nevertheless, the gnostic impulse is far less rare than we might have hoped. The original sin of humankind was a desire to enter the "inner ring" of God, to live not by faith but by gaining access to hidden knowledge ("Ye shall be as gods, *knowing* good and evil"). Adam and Eve were, we might say, the world's first gnostics—and creationists as well as ultra-Darwinists are their sons and daughters.

Whenever we find creationism interwoven with "patriotic" dominion or Christian reconstructionist theologies based upon notions of needing to restore the doctrinal purity of a mythically lost "Christian nation" that has been corrupted by a cabal of evil unbelievers, evolutionists and "secular humanists"; whenever we witness last generation theologies according to which the Messiah cannot return until a certain number of people attain moral and spiritual perfection that includes unwavering confidence in young earth creationism; whenever we see honest questions about faith and science being silenced and committed Christians being branded deviationists or "infidels" for not holding fast to "scientific" creationism as the seal of what it means to be part of God's true "remnant"; whenever we find creationism being subtly or overtly linked with conspiracy theories about the secret agendas of allegedly nefarious political and religious institutions (the Catholic Church, the Freemasons, etc.); whenever we hear that the evidence for "scientific" creationism is to be found not in the weight of the empirical evidence but in esoteric places (hieroglyphics from the tombs of the pharaohs, petrified wood from Noah's ark glimpsed by intrepid Christian explorers on Mount Ararat, clues in Plato's dialogues to the lost City of Atlantis, etc.)—in all such cases we are faced not simply with particularly dogmatic, feverish or fanciful forms of religious imagination but with striking examples of the persistence through the ages of the gnostic temptation.

8

Four Witnesses on the Literal Meaning of Genesis

Barth, Calvin, Augustine and Maimonides

Many literalist readers of the creation narratives in Genesis are not, however, rigid, dogmatic or fundamentalist believers in the least but rather individuals of sincere religious commitment and spiritual depth to whom we owe a debt of gratitude for their contributions to Christian life. They are our friends and family members who have nurtured us in the life of faith, and many of us would not be believers of any kind were it not for them. I think, for example, of my grandparents, who read the Bible in a highly literal way, but who were also people of great generosity, gentleness, courtesy and openness who would never have sought to drive others (certainly not their grandson!) out of their church for the fact that they interpreted Genesis 1 differently from them. Much of the conflict among Christians over questions of biblical authority and evolutionary science, I am convinced, is not propositional or scientific in the final analysis so much as deeply personal. This is provoked in no small part by the stifling demand for mental *sameness* on the part of literalists—the compulsive desire we all at times feel to form those closest to us in our own images—but also by the condescension of people with higher levels of education toward the "simple" faiths of people with less. I wince when I read Conor Cunningham speaking of creationism purely and simply as a "lapse into intellectual barbarism" and "complete desertion of the Christian tradition."[1]

At one level of analysis, I have already made clear, I agree with Cunningham's critique of creationist thinking. At its best, biblical literalism on Genesis appears to me to be a reification of Christian truth and theological expression of what Alfred North Whitehead called *the fallacy of misplaced concreteness*. Yet the creationist family members with whom I eat my Christmas and Thanksgiving dinners are not intellectual or spiritual barbarians. They are often far more charitable, patient, civil and knowledgeable about biblical and theological matters than I. We must carefully distinguish, then, between two distinct mentalities and approaches to the text: literal readings vs. literal*ism*. There are believers on both sides of this divide, and the difference matters. To illustrate this fact, I shall enlist the help of four towering witnesses of the Jewish and Christian faiths whose writings demonstrate the rich variety of ways believers throughout history have thought about the *literal* meaning of Genesis.

In a letter to his niece Christine, the great twentieth-century Swiss theologian Karl Barth wrote:

> The creation story deals only . . . with the revelation of God, which is inaccessible to science as such. The theory of evolution deals with what has become, as it appears to human observation and research and as it invites human interpretation. Thus one's attitude to the creation story and the theory of evolution can take the form of an either/or only if one shuts oneself off completely from faith in God's revelation or from the opportunity for scientific understanding.[2]

What was it, we must ask, that led Barth to describe biblical literalism's approach to the creation not simply as incorrect but as shutting us off *completely* from faith in God's Word?

Barth's key hermeneutical insight is that Scripture, in the deepest sense of the word, is *not* the revelation of God. This may at first sound like a shocking thing for a theologian (described by Pope Pius XII as the greatest Christian thinker since Thomas Aquinas) to say. But for Barth, believers must never forget that there is always something or Someone to which the words of Scripture point. Scripture, in the language of modern semiotics, is the signifier, *not* the Signified. The biblical text is thus not God's revelation but an authoritative *witness to* the revelation of God, which for Christians is God incarnate, the person of Jesus Christ.

Literalism assumes, however, that human language is intrinsically capable not only of guiding us toward God but also of unambiguously revealing realities of God in the form of linguistic signifiers. What is more, it confidently declares that we can read these signifiers without any necessary gap between what the signifying text says and who God really is in his dealings with his creation. In the name of honoring God's Word, literalism thus subtly displaces God's actual sovereignty. It tends toward a kind of idolatry—what we might call "bibliolatry"—that stresses propositional truths and human cognitive powers in a way that denies the radical otherness of God and the freedom of the Creator to *be* the Creator beyond all human signifying speech, including even the speech of the Bible itself.

However, there is an opposite and equally dangerous error that many "liberal" Christians fall into, which Barth referred to as the error of *expressivism*. Expressivism declares that theological language is purely symbolic or "mythical," based upon noncognitive or emotive experiences without any reference to God's actions in human or natural history. According to this view, the Bible tells us only about humanity rather than about God himself and so can only be approached through modern tools of historical-critical scholarship. While uncovering the ways people have grasped for God through history, it denies the fact that God has indeed *entered* history. Expressivism fails to acknowledge God's desire to be known. God has taken the initiative. God has graciously given us his Word. "Whereas literalism underestimates the mystery of God's otherness," writes George Hunsinger, "expressivism underestimates the miracle of God's self-revelation."[3] We are able to speak of the Bible as God's Word because, by a miracle of grace, God overcomes our intrinsic human and linguistic incapacities.

So where do these statements leave us? Clearly in a place of unresolved tension and paradox. Instead of attempting to eliminate the tension and force epistemological and semantic closure on the text, however, Barth suggests that we need to learn how to live in this place of ambiguity and unanswered questions. We need to learn how to read Scripture *dialectically, analogically* and *Christologically,* and in conversation with other Christians, including those alive today and those who have thought deeply about the Bible through the centuries on whose shoulders we stand. We must maintain a constant awareness both of God's self-revelation in the Bible, and of our own limitations in approaching God's Word. The creation narrative in Genesis, Barth suggested, is best seen as a form of what he called "creation saga," as distinct from both "myth" and "history" alike. Genesis, in Barth's thinking, "really speaks of God's creation and it really speaks prehistorically." It is "the direct opposite of myth" and so must be read literally, yet with a clear awareness that the entire story has assumed the form of poetry rather than of historical chronicle.[4] Put another way, we need to be absolutely faithful to the words of Scripture. We need to read them as *literally* true because they are the words God has given us. We must not subtract one thing. But we must at the same time refuse to *add* one thing to Scripture. When fervent believers declare that the days of Genesis can only possibly mean "literal, contiguous, 24-hour periods identical in time to what we now experience as a week," they are adding words to Scripture (and so may well be placing themselves under the stern judgment of Revelation 22:18). The question literalists must continually ask themselves after Barth is: Are we in danger of transforming our literal reading of the biblical witness into the literal*ism* of fallen and finite human minds?

What would it mean to read Genesis in community with others, including the great cloud of witnesses without whom the Christian church as we know it would not exist? One of the conceits of rigidly literalistic approaches to Genesis is that this thoroughly modern hermeneutic is the only approach in keeping with the great watchwords of the Protestant Reformation: *sola scriptura* and *sola fides*. John Calvin's name is especially dear to many creationists, since he believed in a six-day creation occurring in the recent past. In his commentary on Genesis published in 1554, he strenuously rejected the allegorical methods of the early church father Origen, which he said resulted in the meaning of Scripture being "indiscriminately interpreted" and positively "mangled." Calvin also opposed the Augustinian view that the creation had occurred instantaneously and was only narrated in Genesis as filling six days in order to accommodate limited human minds. It "is too violent a cavil to contend that Moses distributes the work which God perfected at once into six days, for the mere purpose of conveying instruction," Calvin wrote. "Let us rather conclude that God himself took the space of six days, for the purpose of accommodating his works to the capacity of men."[5] Although "one moment is as a thousand years" for God, and although God had no "need of this succession of time," he nevertheless spent six literal days creating the world so "that he might engage us in the consideration of his deeds."[6]

Some readers (for example, Peter M. van Bemmelen in a 2001 article titled "Divine Accommodation and Biblical Creation"[7]) have gone no further than statements such as these to assert that Calvin's importance for contemporary debates over faith and science is his affirmation of a biblical literalism thoroughly consistent with modern creationism. Careful readers of Calvin's commentary on Genesis will find, however, that he held far more complex, intriguing and contemporary-sounding views on Genesis and scientific reasoning than literalists allow.

Calvin believed that all matter was created by God on the first day of creation, but that the plants and animals were *not* created *ex nihilo*. They were formed by God from a welter of preexisting materials and so were organically and materially related both to each other and to the created universe as a whole. Readers who insisted on completely distinct creations out of nothing on each of the days within the creation week were engaged in intellectual sophistry, in Calvin's view. The word *created*, he continued, ought not to be read in an overly literal way as far as questions of chronology are concerned; creatures were, in effect, "created" before they were *formed*:

> Those who assert that the fishes were created from nothing because the waters were in no way sufficient or suitable for their production are nevertheless resorting to rationalization, for the fact would remain that the material of which they were made existed before, which, strictly speaking the word "created" does not admit. I therefore do not restrict the creation here spoken of to the work of the fifth day but rather suppose it to refer to that shapeless and confused mass that was in effect the fountain of the whole world.[8]

Calvin's language is difficult, but he clearly rejected the idea that the creation in Genesis only describes events and not processes. He seems to have held a kind of hyperaccelerated emergent or even evolutionary view of what happened during the creation week:

> [God] created the great creatures of the sea and other fishes—not that the beginning of their creation is to be reckoned from the moment in which they received their form but because of the universal matter that was made out of nothing. So with respect to species, form only was added to them; but creation is nevertheless a term truly used respecting both the whole and the parts.[9]

The reason God created the animals using "material from the earth" was not "because he needed it, but in order that he might combine the separate parts of the world with the universe itself."[10] While declaring that the creatures were formed by God not *ex nihilo* but out of "that shapeless and confused mass" that was "the fountain of the whole world," Calvin nevertheless declared that the imparting of life to matter was a miraculous, scientifically unexplainable event. "From where does a dead element gain life? This is in this respect a miracle as great as if God had begun to create out of nothing those things that he commanded to proceed from the earth."[11]

What might Calvin say, then, about highly literalistic interpretations of

Genesis today in the light of scientific evidences for a very old earth and common ancestry across diverse species? The clearest evidence is given by Calvin in his interpretation of Genesis 1:16. Calvin's concern was to make clear that the language of "greater" and "lesser" lights in Genesis in no way conflicted with contemporary astronomical calculations that showed that seemingly small stars and planets such as Saturn were actually much greater in size than the "great" light of the moon ordained by God to "govern the night." While this might seem like a trivial matter to us today, it was an unsettling challenge to the biblically inspired and geocentric cosmological theories of medieval and early Reformation Europe. Calvin's approach to the problem is simple: *the words of Genesis are not to be taken literally.* The Reformer's clearest statement on the relationship between the scientific evidence of his day and the theological meaning of Genesis is as follows (the passage deserves to be quoted at length):

> Moses described in popular style what all ordinary men without training and education perceive with their ordinary senses. Astronomers, on the other hand, investigate with great labor whatever the keenness of man's intellect is able to discover. Such study is certainly not to be disapproved, nor science condemned with the insolence of some fanatics who habitually reject whatever is unknown to them. . . .
>
> Moses did not wish to keep us from such study when he omitted the scientific details. But since he had been appointed a guide of unlearned men rather than of the learned, he could not fulfill his duty except by coming down to their level. If he had spoken of matters unknown to the crowd, the unlearned could say that his teaching was over their heads. In fact, when the spirit of God opens a common school for all, it is not strange that he chooses to teach especially what can be understood by all.
>
> When the astronomer seeks the true size of stars and finds the moon smaller than Saturn, he gives us specialized knowledge. But the eye sees things differently, and Moses adapts himself to the ordinary view.

> God has stretched out his hand to us to give us the splendor of the sun and moon to enjoy. Great would be our ingratitude if we shut our eyes to this experience of beauty! There is no reason why clever men should jeer at Moses' ignorance. He is not explaining the heavens to us but is describing what is before our eyes. Let the astronomers possess their own deeper knowledge. Meanwhile, those who see the nightly splendor of the moon are possessed by perverse ingratitude if they do not recognize the goodness of God.[12]

In short, the great leader of the Reformation urged a flexible and nondogmatic approach to questions of origins, cosmology and science that assumed: (1) that the Genesis writer "omitted the scientific details"; (2) that believers should not protest like "fanatics" if scientists "possess their own deeper knowledge"; (3) that where the creation narrative conflicts with new scientific evidence it should not be read literally "nor science condemned"; and (4) that the purpose of Genesis was not to describe scientific facts but to open our eyes to the "experience of beauty" and to "the goodness of God."

Unfortunately, there are other highly disturbing facts about Calvin's life and theology that might also be of some tragic relevance to contemporary debates over science and the meaning of Genesis. Calvin did model a constructive and flexible epistemology and hermeneutic in response to new astronomical data that challenged longstanding understandings of Genesis 1. He failed, however, to consistently exemplify virtues of intellectual openness, toleration and respect for freedom of conscience within the body of Christ. Calvin's complicity in the 1553 trial and execution of the Spanish Anabaptist Michael Servetus, who denied the doctrine of the Trinity and the practice of infant baptism, must forever stand as a warning tale of the calamities that have historically followed attempts by those in positions of ecclesial power to forge unity in the body of Christ by imposing doctrinal *uniformity* upon others within their communities at any cost. In his 1554 *Defensio orthodoxae fidei de sacra Trinitate,* Calvin offered the following theological rationale for religious persecution of unregenerate "heretics" and defense of his own role in the trial of Servetus: "[Civil and religious authority] exists so that idolatry, sacrilege of the name of God, blasphemies against his truth and other public offenses against religion may not emerge and be disseminated."[13]

During the proceedings, Calvin displayed his relative "moderation" by arguing that Servetus be given the punishment of beheading (normally reserved for crimes of sedition) rather than burning (the prescribed penalty for heresy). The court, however, saw no reason to mitigate Servetus's sentence, which by its nature would serve as a vivid, edifying lesson to the entire community of the fate awaiting all unbelievers. As the Christian humanist and champion of religious liberty was being burned alive on the plain of Champel outside the gate of Geneva, he reportedly cried out, "Oh Jesus, son of eternal God, have pity on me!" His words provided the zealous citizens of Geneva with final confirmation of his guilt, since the proper trinitarian cry of dereliction would have been: Oh Jesus, *eternal son of God*, have pity on me. "His punishment was due to the misplacing of a single adjective," Calvin biographer Bernard Cottret writes of Servetus. "Heresy is never anything but a question of grammar."[14]

More than a millennium before Calvin wrote his commentary on Genesis, several church fathers expressed similarly open views about the relationship between science and biblical authority. In his treatise on the creation, *The Literal Meaning of Genesis,* Augustine of Hippo (A.D. 354–430) argued against literalistic readings of Genesis precisely in the "scientific" sense today's creationists demand. Calvin and Augustine in fact agree far more than they disagree on how we ought to think about Genesis and scientific knowledge. Although Augustine had no awareness of the theory of evolution, here is what he had to say about the relationship between Scripture and the cosmological evidences available in his own time:

> Usually, even a non-Christian knows something about the earth, the heavens, and the other elements of this world, about the motion and orbit of the stars and even their size and relative positions,

about the predictable eclipses of the sun and moon, the cycles of the years and the seasons, about the kinds of animals, shrubs, stones, and so forth, and this knowledge he holds to as being certain from reason and experience. Now, it is a disgraceful and dangerous thing for an infidel to hear a Christian, presumably giving the meaning of Holy Scripture, talking nonsense on these topics; and we should take all means to prevent such an embarrassing situation, in which people show up vast ignorance in a Christian and laugh it to scorn. The shame is not so much that an ignorant individual is derided, but that people outside the household of faith think our sacred writers held such opinions, and, to the great loss of those for whose salvation we toil, the writers of our Scripture are criticized and rejected as unlearned men.

If they find a Christian mistaken in a field which they themselves know well and hear him maintaining his foolish opinions about our books, how are they going to believe those books in matters concerning the resurrection of the dead, the hope of eternal life, and the kingdom of heaven, when they think their pages are full of falsehoods and on facts which they themselves have learnt from experience and the light of reason? Reckless and incompetent expounders of Holy Scripture bring untold trouble and sorrow on their wiser brethren when they are caught in one of their mischievous false opinions and are taken to task by those who are not bound by the authority of our sacred books. For then, to defend their utterly foolish and obviously untrue statements, they will try to call upon Holy Scripture for proof and even recite from memory many passages which they think support their position, although they understand neither what they say nor the things about which they make assertion.[15]

In his 2009 Gifford Lectures at the University of Aberdeen, published under the title *A Fine-Tuned Universe,* Alister McGrath (a convert to Christianity from atheism, a distinguished theologian and church historian, and an Oxford-trained biologist who has publicly debated the likes of Richard Dawkins and Daniel Dennett on questions of religion and science) finds vital resources in Augustine's theology for addressing contemporary questions of origins and evolutionary biology. McGrath draws in particular on Augustine's concept of *rationes seminales,* or "seminal reasons."

In Augustine's thought, God's great wisdom and creative power are revealed not only in his ability to create *ex nihilo* but also in his ability to impart seedlike potentialities or "embedded causalities" (as McGrath calls them) into the creation. These *rationes seminales* unfold over time so that the creation must be understood as both process and event. McGrath employs these ideas of Augustine's in defense of an evolutionary understanding of creation that embraces Darwinian ideas while arguing there is far more to the story than Darwin imagined and than can be analyzed at the level of material causation alone.

At the same time, McGrath cautions, Augustine's thought has clear weaknesses that reflect the time and place from which he wrote. For one thing, Augustine's vision of the natural world is far more static than modern biology supports. In Augustine's thinking, the process of creation—that is, the divinely orchestrated emergence of new forms from the mutable "seeds" of the original creation event—leads to fixed or immutable biological forms. Augustine was led astray, McGrath suggests, not by Scripture but by his commitment to the Aristotelian teleology of his day. Aristotle's highly essentialist view of species became so deeply embedded in Christian thinking, McGrath provocatively writes, "that [in the nineteenth and twentieth centuries] a scientific challenge to Aristotle [namely the theory of natural selection] was misread as a scientific challenge to the Bible."[16] However, "It was the lens through which Scripture was read, not Scripture itself, that was challenged by Darwin's notion of the evolution of species."[17]

In addition to these and other witnesses from the Christian tradition, biblical literalists would do well to heed the nonliteralistic readings of Genesis spanning centuries if not millennia within the Jewish faith. Orthodox Judaism has not been disturbed by evolutionary concepts or obsessed with "scientific" creationism the way that conservative Christians have been. Although some ultra-Orthodox Jewish sects have denied the reality of dinosaurs and torn the pages dealing with evolution out of their textbooks, Ira Robinson notes, in general "fundamentalism and creationism have been decried by Orthodox Jews as 'nonsense' and 'a grave error'"; even if Genesis is read in a highly literal manner, the biblical texts in broad Jewish perspective *literally* do not mean what Christian creationists say they mean.[18] The Hebrew Bible of course also belongs to the Jewish faith, and it did so long before the arrival of Christianity. Christians might therefore have much to learn from Jewish interpretive traditions.

Perhaps the most revered and authoritative interpreter of Hebrew Scripture in Jewish history is the rabbi, physician and philosopher Moses Maimonides. Maimonides was born in Cordoba, Spain, in 1135 but was forced to flee with his family at an early age to Morocco before finally settling in Egypt when Spain was invaded by the Almohad caliphate. According to a popular medieval Jewish saying repeated by Orthodox Jews up to the present, "From Moses [in the Torah] to Moses [Maimonides] there was none like Moses."[19] Some Christians are casually dismissive of Maimonides, along with other great medieval thinkers such as Aquinas (who was in fact heavily indebted to Maimonides), claiming that their theology has nothing to teach us because it was corrupted by a syncretistic accommodation of pagan Greek philosophy. In fact, one of Maimonides's major concerns in his masterpiece, *The Guide for the Perplexed* (completed in 1190), was to demonstrate, in opposition to both Plato and Aristotle, that the creation occurred as the Hebrew Bible records it.[20] In Plato's *Timaeus*, written around the middle of the fourth century B.C.E., the world is created by a demiurge *de novo* (that is, at a moment in time) but not *ex nihilo* (out of nothing). Plato imagines a creation from preexisting matter in the cosmos. For Aristotle, by contrast, the world was created neither *de novo* nor *ex nihilo*. It is instead an eternal emanation of the Unmoved Mover without beginning or end. Although Maimonides confessed that he could not disprove the logical coherence of the Aristotelian view, he insisted from the revealed authority of Scripture that the creation occurred both *de novo* and *ex nihilo*.

> [A]ll that Aristotle and his followers have set forth in the way of proof of the eternity of the world does not constitute in my opinion a cogent demonstration, but rather arguments subject to grave doubts. . . . What I myself desire to make clear is that the world's being created in time, according to the opinion of our Law . . . is not impossible and that all those philosophical proofs from which it seems that the matter is different from what we have stated, all those arguments have a certain point through which they may be invalidated and the inference drawn from them against us shown to be incorrect.[21]

Even as Maimonides argued for the superiority of Hebrew Scripture over Greek philosophy on important questions of origins, he also declared that it was a mistake to read the six days of the Genesis narrative as literal twenty-four-hour periods. The creation narratives in Genesis, Maimonides taught, should not be understood as a *cosmogony,* that is, as a detailed scientific or historical description of the way the world came into being. Instead, they should be grasped as a *cosmology*—as a description of the structure and order of God's creation and the way the parts fit with the whole. Strictly speaking, the question of how the world first came into being is undiscoverable by scientific methods and remains veiled in mystery. This mystery persists even within the biblical narrative itself, although Scripture tells us what is most important to know about God's care for earth and for all living creatures. The theological meaning of Genesis for Maimonides was not tied, then, to any kind of unbending or chronological literalism. The narratives of Genesis 1–2 are theological-metaphorical—not historical or "scientific," nor allegorical[i]—in nature.

---

[i]The allegorical approach to reading Genesis is most strongly associated in the Jewish tradition with the Hellenizing writings of Philo of Alexandria, who was born some twenty-five years before Christ and who ultimately had a far greater influence on Christian interpretations of Scripture than Jewish ones. The extent to which allegorical approaches to different parts of the Hebrew Bible, including the Pentateuch, were embraced by Palestinian rabbis

Maimonides arrived at this reading of Genesis, it is important to note, under no pressure to conform his theology to new scientific discoveries or evolutionary theories. However, he taught that, in cases where the weight of the scientific evidence *is* clear, believers should not hesitate to modify their interpretations of Scripture as reason and new discoveries might lead. Scientific findings should not be denied nor bent every which way to somehow fit one's prior assumptions about the meanings of Scripture. Instead, the rabbi's guiding principle was that "one should accept the truth from whatever source it proceeds."[22] There can be no conflict between scientific reasoning and correct interpretations of the Bible, Maimonides maintained. Hence, if one finds that something can be proven beyond reasonable doubt by scientific methods but that it *seems* to conflict with Scripture, this means one has misinterpreted Scripture. If, on the other hand, a statement is made that is rationally plausible but is in principle unprovable by science, *and* if this statement conflicts with the clear teachings of Scripture (such as Aristotle's conception of the cosmos as an eternally existing emanation without any beginning), the believer should reject the statement as both theologically and scientifically unsound.

For Maimonides, the Protestant slogan of *sola scriptura* might therefore be well and good as far as it goes as a statement of Scripture's supreme authority in theological matters. But those who abandon their reason, their observations and their senses when *interpreting* the Bible are in fact not showing any great honor to the Bible's authority. They are actually distorting Scripture's authentic meaning and sealing themselves off to divine revelation by refusing to use their God-given minds to discern revelation's truths. The most faithful way of reading Scripture includes leaving space for

---

in the first century B.C.E. even before Philo is a matter of scholarly debate. By all indications Genesis was read in a quite literalistic fashion by most if not all Palestinian Jews during this period. But it is also clear that the tradition was a constantly evolving rather than static one, and that it included hermeneutical resources and precedents for nonliteralistic readings to emerge. Strict literalism on the days of Genesis was never thought of in either ancient Judaism or early Christianity as a kind of foundationalist litmus test for Jewish or Christian identity.

scientific reasoning as a *corrective* to one's "commonsense" intuitions or subjective readings, which would otherwise be prone to solipsism and error. Put another way, science is not simply to be treated by the believer as an instrumental tool for rationalistic apologetics but rather as a vital aid to spiritual discernment. "One can easily fall prey to the illusion that one understands Scripture by virtue of being able to read Hebrew," Gad Freudenthal writes of Maimonides's hermeneutics. "In point of fact, many words and phrases are, as it were, encoded—they have a particular, philosophic sense, so that understanding them on their ordinary meaning inevitably leads to error, even heresy. For the naive reader, the revealed text is therefore full of pitfalls."[23]

How, then, might Maimonides respond to the challenges of evolutionary theory if he were alive today? He might begin by drawing vital distinctions between reductionistic forms of Darwinism on the one hand and theistic theories of organic evolution on the other. The two are not the same, much as creationists strive to conflate them (seemingly in order to avoid having to attend to what their fellow believers are actually saying). He would surely provide a vigorous critique—on scientific as well as theological grounds—of the pretensions of the so-called new atheists with their unscientific and philosophically vacuous claims that Darwin has somehow proven that there is no purpose, no meaning and no divine providence in natural history. At the same time, I would venture, Maimonides along with most Jews today would embrace with perfect equanimity falsifiable scientific tests demonstrating beyond reasonable doubt that the earth is very old, that there has been life on earth for a very long time, and that there were predatory creatures on earth before the appearance of human beings.

We should not end this discussion of Maimonides, though, without recalling the controversy that surrounded him during his lifetime and for some time after. In medieval Judaism, unlike in medieval Christianity, the charge of heresy was extremely rare. Jews did not traditionally define membership in the community or devotion to God in terms of one's mental assent to a fixed set of propositional formulas. They increasingly came to do so, however, under the pressure of their surrounding Christian cultures, which often violently demanded that Jews indicate what specific doctrinal formulations they did and did not accept. Maimonides himself played an important role in this development by compiling a (much resisted) list of thirteen beliefs from Talmudic sources that he said one must embrace to be a devout Jew.[24]

Some Jewish communities were so perplexed by Maimonides's *Guide for the Perplexed* that they banned the book and publicly burned it. These attacks on the rabbi had as much to do with political and sociological realities as they did with doctrinal disagreements. Leaders in the rabbinic academies—who viewed themselves as the sole authoritative interpreters of the Torah—saw Maimonides's writings as a threat to their institutional power. Ashkenazi Jews from northern France meanwhile worried that Judaism was being corrupted by secular philosophical learning and clashed vehemently with Sephardic Jews from Andalusia, who drew freely upon the rich traditions of both Arabic and Greek learning.[25] Rabbi Solomon ben Abraham, representing the "conservative" Ashkenazi Jews, imposed a ban on Maimonides—to which the Sephardic rabbis of Lunel simply proclaimed a counterban. These debates spilled into the Catholic Church, with the Inquisition eventually also burning Maimonides's books. On March 7, 1277, some seventy years after Maimonides's death, the Bishop of Paris, Stephen Tempier, famously banned the teaching of 219 philosophical and theological theses in an attempt to purge the theology faculty at the University of Paris of what he deemed pagan learning. The unnamed targets of this campaign of censorship (probably inspired by conservative Catholic theologians and clerics at the university who had the ear of the bishop) included Aquinas, Averroës and Maimonides, the three most influential thinkers to have reintroduced Aristotle's philosophy to the West.

Maimonides, though, was undeterred by the religious zealots who sought to suppress his writings during his lifetime. Their reward was their own confusion. They were not, he wrote, the honest seekers of truth to whom he addressed his commentaries:

I am the man who when the concern pressed him and his way was straitened and he could find no other device by which to teach a demonstrated truth other than by giving satisfaction to a single virtuous man while displeasing ten thousand ignoramuses—I am he who prefers to address that single man by himself, and I do not heed the blame of those many creatures.[26]

9

If Not Foundationalism, What Then?

From Tower Building to Net Mending

"**Much philosophical confusion**," Nancey Murphy writes (with reference to Wittgenstein), "comes from being captivated by a picture."[1] The simplifying stories, images, analogies, metaphors and allusions we use to represent our ideas to others can easily come to substitute for clear thinking and serve to insulate our ideas from scrutiny, even by ourselves. For example, I have heard not one but several creationists speak of the moral responsibilities of Christian educators in terms of a dubious mental picture drawn straight from the world of corporate America. If you work for Nike but spend your office hours promoting Reebok as the better shoemaker, these individuals declare, you are in fact *stealing* from your employer. Hence, professors who teach evolutionary concepts at Christian colleges and universities whose sponsoring denominations officially adhere to creationism are not merely theologically mistaken but are morally despicable. They are, in the words of a televangelist I once watched, breaking the eighth commandment ("Thou shalt not steal") and so are "falling under the condemnation of God." (The speaker, I later learned, had in fact dropped out of college, such was the depth of his commitment to Christian education; as Flannery O'Connor wrote, "Conviction without experience makes for harshness."[2])

But why, we must ask, did this individual assume to begin with that the body of Christ is best thought of in terms of the picture of a multinational corporation aimed at maximizing profits for investors, with the search for truth evidently conceived as brand loyalty in a zero-sum competition for people's souls or minds as currency or capital? What sorts of hidden values, priorities and assumptions are at work when the rules of corporate capitalism provide the controlling metaphors for preachers explaining to their flocks the meaning of church unity, intellectual integrity, moral decision making and Christian discipleship?

When very poor pictures such as this are deployed as substitutes for careful thought, the best response one can give is to suggest an alternative picture (metaphor or analogy) that might help to reorient our horizons. For example, what if Christian universities and Christian classrooms are best thought of not as shoe factories (with teachers in the role of wage-laborers paid to assemble "products" with maximum efficiency according to the specifications laid down by CEOs), but instead as courts—that is, as deliberative or judicial assemblies charged with wrestling with difficult questions of truth and justice for the good of the community as a whole?

Unanimous Supreme Court decisions are rare. Practically every US Supreme Court ruling on every major issue includes one or more dissenting opinion(s). These dissenting opinions are clearly and publicly articulated and might in the future influence the overturning of an earlier decision. The health of a democratic polity that is oriented toward questions of truth and justice, the framers of the American legal system understood, depends not only on consensus but also on *dissent*. And a dissenting judge is not being "unpatriotic" or defying the law by disagreeing with the majority opinion. They are in fact upholding the deepest meaning of the law in the very act of raising principled objections to it. So here is a question we might ask those who have become convinced that institutions of higher education are corrupting the youth: What if Christian colleges and universities—even those affiliated with traditions with highly literalistic doctrines of creation—embraced a picture of unity in the body of Christ that included the concept of necessary loyal dissent within a framework of basic respect, transparency and honest searching for truth? Communities that instead strive to model their inner workings on pictures of corporate power and control will in the end come to resemble not successful shoemakers so much as oppressive authoritarian regimes.

According to some sociologists of religion, it is impossible for fundamentalist groups to allow for any significantly divergent expressions of belief on a matter like creationism. "Addicted as it is to homogeneity and to equality in purity, a 'loyal opposition' has no place there," Gabriel Almond, R. Scott Appleby and Emmanuel Sivan write. "Any opposition is bound to be accused of treachery and become the object of witch-hunts and ostracism, liable ultimately to be 'excised.'"[3] One test for whether a religious tradition has cast its lot irreversibly with fundamentalisms the world over is whether it sees its doctrinal formulations—however literalistic or nonliteralistic in wording these formulations may be—as present statements of group consensus subject to ongoing dialogue and possible change in the future, or rather as unquestionable, infallible or inerrant dogmas whose purpose is to sharply define insiders and outsiders.

The foundationalist picture of knowledge as a stacked brick tower (which I described in chapter two) is another image that must be challenged with better pictures of the life of the mind. The edifice of philosophical foundationalism is a faulty tower, and to the extent that some Christians have made the task of constructing it definitional for the life of faith it has also proven to be an idolatrous and oppressive tower, resulting only in confusion and broken community. The high attrition of young adults from fundamentalist-leaning denominations (I speak now of many close friends) is a scandal that cannot simply be laid at the feet of the biology professors who introduced these students to the evidence for organic evolution.[4] There comes a point at which the leaders of highly conservative faith communities must ask themselves how many more of their sons and daughters they are prepared to see walk out the doors of their churches never to return again because these young people find no room for intellectual growth, intellectual honesty or openness to new ways of thinking within their community's walls. In the meantime the elders continue to respond to all outside challenges with the tactic of circling the wagons and fixing their foundationalist bayonets (rusted with age, dull as butter knives from overuse).

The alternative to the foundationalist-style apologetics that underlie virtually all versions of young earth or young life creationism is not the great shibboleth of postmodern relativism and nihilism, as some literalists claim. If anything, the shoe is on the other foot. "A nihilist is not one who believes in nothing," Albert Camus wrote, "but one who does not believe in what exists."[5] It is "scientific" creationists themselves who are most clearly beguiled by the postmodern claim that the weight of the empirical evidence when it comes to questions of origins should be irrelevant to what we believe; that scientific findings can be dispensed with as nothing more than a social construct whenever they do not support our prior religious convictions or biblical interpretations; and that obstinate material realities may be reinterpreted virtually at will to fit one's subjective worldview. Evolutionary theory, many creationists tell us, is nothing more than an inverted faith in atheistic materialism or naturalism. "[W]e are forced to choose and in practice accept one of the competing theories [creationism or evolutionary science] as absolutely true," declares theologian Fernando Canale. "This acceptance is not based on reason or method, but on faith or the relative confidence we personally place on the theory we adopt as being the most persuasive explanation of reality."[6]

Canale builds his case upon the work of postmodern philosophers who have challenged Enlightenment notions of universal reason. There is nothing postmodern, however, in his insistence that we embrace a single paradigm of origins as "absolutely true" to the radical exclusion of other legitimate sources of knowledge. This is a thoroughly modernist and foundationalist way of thinking. Canale, recalling Descartes, rules out the possibility of true epistemological holism in favor of a ranked ordering of different sources of knowledge beginning from an indubitable base of absolute certitude. His sophisticated analysis and critique of the socially constructed nature of scientific knowledge fails to adequately wrestle with the other side of the equation: the socially constructed nature of literalism as a fallible human interpretation of Scripture that is in conflict not only with the paradigm of modern science but with the work of a great many theologians and biblical scholars—persons of deep religious commitment who argue on compelling grounds that the Hebrew Bible was never meant to be read the way contemporary creationists read it. Still, is Canale right that what postmodernism has taught us is that at the end of the day evolutionary science is really just an alternative "metanarrative" that is "not based on reason or method" but on a leap of faith that plunges us into a perhaps internally consistent paradigm that is nevertheless completely incompatible with the Christian worldview?

All that really matters, according to this way of thinking, are one's presuppositions. Simply change your conceptual lens, and the geological columns may not exactly rearrange themselves but will at least appear in a new light, suddenly congenial to the literalist's will to believe. We should be radically skeptical postmodern relativists and social constructivists when it comes to inconvenient scientific and historical truths, it seems . . . but thoroughgoing modernists committed to a strictly foundationalist epistemology, a literalistic hermeneutic, a fundamentalist ecclesiology and a "scientific" apologetic the moment we open our Bibles. Is it any wonder that so many young adults who are told that dancing this awkward two-step is the only way to remain believers in the modern world decide to sit the dance out—or to abandon religious life entirely—when the cognitive dissonance becomes too great?

In answer to the epistemological relativism if not nihilism of "scientific" creationism (that emerges as an ironic consequence of the literalist's quest for a single source of absolute, indubitable certainty, reminding us of how the postmodern turn was incipient in the modernist project from its start), an increasing number of evangelical and other Christian scholars have advanced a picture of knowledge as a form of *critical realism* and *postfoundationalism.*

Critical realism seeks to chart a third way between the epistemologies of both modernism and postmodernism. It is *critical* because it accepts the postmodern emphasis on the provisional and always contingent or mediated aspect of knowledge, yet *realist* because it insists upon the objectivity of the world we encounter and so the possibility of more or less truthful ways of talking about the properties of this world. Whether in matters of science *or* biblical interpretation we must resist the temptation of modernist positivism, "commonsense empiricism," or naive realism (the dubious claim that we have direct, unmediated or unproblematic access to truth). At the same time, we must resist the temptation of sheer postmodern relativism (the claim that truth or even reality as such does not exist or cannot be known at all apart from purely subjective language games about it). There is a world of glorious, messy, complex and stubborn biological and physical facts that are real, that we have not created and to which we must be intellectually accountable lest we deny God's actual creation in the name of an anxious and solipsistic creation*ism.*

One eloquent spokesman for critical realism is New Testament scholar N. T. Wright. "If knowing something is like looking through a telescope," he says,

> a simplistic positivist [or modernist] might imagine he is simply looking at the object, forgetting for the moment the fact that he is looking through lenses, while a phenomenalist [or postmodernist] might suspect that she is looking at a mirror, in which she is seeing the reflection of her own eye.[7]

We never possess a "god's-eye view" of the world—a view without lenses, so to speak—but neither are we simply trapped in a hall of mirrors. We must be critically responsive to both biblical *and* scientific evidence "along the spiraling path of *appropriate dialogue or conversation between the knower and the thing known*" (emphasis original).[8] All of us, Wright points out, inevitably interpret the information we receive, no matter the source, through a complex grid of memories, narratives, cultural expectations and psychological needs or desires. We are unavoidably shaped by the traditions we are a part of, and we often tenaciously cling to our beliefs for precisely this reason—not because they are the best or only fit to the evidence but because they are part of the story we have long told about ourselves, and we feel that giving up or altering part of our story would mean giving up our entire identity. "Every human community shares and cherishes certain assumptions, traditions, expectations, and anxieties, and so forth, which encourage its members to construe reality in particular ways, and which create contexts within which certain kinds of statements are perceived as making sense."[9] What all this means is that we must abandon any pretensions to absolute certainty about our interpretations and confess that we always see "through a glass darkly" (1 Cor 13:12). This is certainly true in scientific matters but is perhaps *especially* true in theological ones (it was, after all, theology and not science that the apostle Paul was speaking of when he used the "through a glass darkly" metaphor).

Another implication of critical realism is that we must always see our knowledge about particular realities as fitting within narrative frameworks that are nested within larger stories and ultimately within worldviews that involve complex interplays of reason, observation, culture, history and experience that cannot be understood in any kind of neatly stacked way. We must chart a course between the Scylla of modernist-style foundationalism and the Charybdis of postmodern *anti*foundationalism. The word for this post-postmodern position is *post*foundationalism (as if the prefix "post" were not overused by academics already!). Postfoundationalists agree with foundationalists that our worldviews can have greater or lesser correspondence with reality, including both scientific and theological truth. They agree with antifoundationalists (and nonfoundationalists), though, that the attempt to build a system of knowledge from a base of indubitable, infallible certitude that somehow stands on its own (whether this base is said to be the "plain" words of Scripture or human sensory experiences or something else) is an utterly failed epistemological project well past its sell-by date.

The postfoundationalist picture of knowledge has often been described as a "net" or "web" of truth. The strength and stability of a net does not depend upon any one of its nodes but upon the entire field as a unified whole. We can alter or mend different cross-points in the net using a wide variety of "knots" (methods) without fear that doing so will cause the entire net to come unraveled. Net menders will not be overly troubled by different readings of Genesis or other parts of Scripture so long as these readings preserve the overall coherence of the Christian narrative. To see knowledge as a web or net in this way is to embrace true epistemological holism and interdisciplinary dialogue, for the interlacing strands of a web always run in both directions. The nodes in the net of Christian truth—which include knowledge gleaned from tradition, reason, observation and Scripture—are mutually supportive and mutually defining, never standing alone the way the foundation to a brick tower does. A postfoundationalist and critical realist perspective on both the book of Genesis and the "Book of Nature" allows for a genuine conversation between religion and science to occur in ways that foundationalist epistemologies (whether religious or secular) do not.

Foundationalists, we might say, are unable to fix leaking pipes and shorting power lines buried beneath their homes since to even see these problems they would have to first dismantle the structures they have built on top of them. Postfoundationalists do not have this same difficulty since (to use a simile made famous by the philosopher Otto Neurath) their entire orientation allows them to engage in epistemological and hermeneutical repair work like sailors at sea. If we think of Christian beliefs and practices as the planks that make up a vessel (the church), we can imagine these planks being replaced one by one during a long voyage so that by the end of the journey the entire boat has been transformed into a new ship that is still, paradoxically, the same ship that left port. It would be madness, however, to attempt to replace an entire ship all at once on open waters. The work of repair is only possible because the ship is strong enough to continue sailing while some of its planks are being restored. Dynamic change requires the continuity and stability of the total system, while the stability of the system requires constant dynamic change at the level of its constituent parts.

All metaphors at some point break down, but hopefully these pictures will help readers better grasp my approach to the theodicy dilemma of animal suffering in the remainder of this book. I assume the overall coherence as well as the compatibility of scientific and theological frameworks, the postfoundationalist character of all knowledge, the need for intellectual openness to new insights from diverse sources, and the necessity of developing an understanding of animal suffering consistent with God's character of love as revealed in the historical person of Christ. I believe that Christians can boldly pursue the work of doctrinal "net mending" or "ship repair" in the light of challenging scientific evidence, since all truth is ultimately one. I fear that those who refuse to renovate their ships on the conviction that every plank is foundational to some other plank will in the end find themselves clinging to a single board as their vessels break apart into so much driftwood on the high seas.

Part Two

On Animal Suffering

"In Praise of Self-Deprecation"

The buzzard has nothing to fault himself with.
Scruples are alien to the black panther.
Piranhas do not doubt the rightness of their actions.
The rattlesnake approves of himself without reservations.

The self-critical jackal does not exist.
The locust, alligator, trichina, horsefly
live as they live and are glad of it.

The killer whale's heart weighs one hundred kilos
but in other respects it is light.

There is nothing more animal-like
than a clear conscience
on the third planet of the Sun.

—Wislawa Szymborska

10

Stasis, Deception, Curse

Three Literalist Dilemmas

Let us be clear from the outset that there are no tidy answers to the theodicy dilemma of animal suffering, and that any theology that accepts "deep time" and organic evolution (not to be confused with ultra-Darwinism) as established facts of science is indeed faced with severe difficulties—just as anyone who accepts the Nazi or Rwandan genocides as established facts of history is faced with severe difficulties. The longer I have pondered these challenges, though, the more convinced I have become that they are far less severe than the theodicy dilemma posed by strict literalism or "scientific" creationism. Before attempting any positive response to the problem, we must first see a surprisingly obvious but widely overlooked fact about literalism and creationism in their claims to answer the theodicy riddle posed by animal predation in a more satisfying or morally coherent way than other approaches to Genesis.

What biblical literalists and creationists have collectively failed to grasp is that strictly Darwinian evolutionary accounts do not introduce the theodicy dilemma posed by animal suffering but rather *solve* it, albeit through the elimination of the Subject who makes the problem a moral problem as such. The idea that an omnipotent Creator designed the natural world "revolts our understanding," Darwin wrote in one of his private letters, "for what advantage can there be in the sufferings of millions of lower animals throughout almost endless time?" By contrast, he concluded, "the presence of much suffering agrees well" with "natural selection."[1]

Darwin's framing of the problem of "natural evil" is no less compelling if we shorten the period in question from "almost endless time" to "only" thousands of years. We must confront the magnitude of animal suffering and death occurring every instant of every day, which evolutionary theory did not create but which it at least has the theological advantage of explaining in a way that does not directly attribute the harshest facts of nature to God's wrathful response to human sin. "During the minute it takes me to compose this sentence," Richard Dawkins writes in *River Out of Eden*, "thousands of animals are being eaten alive; others are running for their lives, whimpering with fear; others are being slowly devoured from within by rasping parasites; thousands of all kinds are dying of starvation, thirst and disease."[2] One need not be impressed by Dawkins's philosophically vacuous "new atheism" to appreciate the moral gravity of the problem he identifies here. Atheists must wrestle with a weighty philosophical problem of their own, namely, the "problem" of goodness: why is it that we perceive such deep meaning, beauty and value in so much of the world if at the end of the day it is all just atoms in the void? But it is left to creationists who attribute all of the dysteleological and troubling realities of animal existence to God's "curse" upon the animal kingdom to explain why a fully just, fully loving and omnipotent Creator would not simply permit but positively *demand* such suffering among uncomprehending and morally innocent creatures who were previously unexposed to pain or death of any kind. Although evolutionary frameworks greatly increase the *quantity* of animal suffering by extending its reality across longer stretches of time, the most theologically coherent theodical account is not necessarily the one that yields the shortest time frame. Were this the case, thousands of years of animal and human suffering would already be more than enough to cause any sensitive mind to revolt, for God could surely have put an end to evil the instant it was born.

According to the traditional literalist telling, God created a completely deathless world that only became subject to suffering and mortality as a result of Adam's disobedience (occurring, most creationists believe, within the past six to ten thousand years). Christ's atoning death on the cross has made possible an escape for those humans who by faith accept his sacrifice and so find deliverance from the penalty of sin they have inherited from Adam. But the world of nature in its present form must ultimately be destroyed, and all of its animals with it. After the earth is consumed or cleansed with fire, those humans who have been saved by faith will inhabit an earth made new in which death and pain will be no more, just as it was in the beginning. It is therefore essential to both our soteriology and our eschatology, most biblical literalists assert, that all death and all suffering in the world be attributed to Adam's rebellion alone, otherwise Christ's death as the second Adam would be emptied of its meaning.

Yet expiatory atonement theories such as these that insist all nonhuman mortality must be directly attributed to human sin do not solve the theodicy problem presented by predation and animal suffering in the least. In fact, many believers may be surprised to discover, they raise it to harrowing new heights. It is time that biblical literalists at least candidly acknowledge that the challenges they face are not only scientific but theological and moral as well, and that these problems are no less great for them than for process creationists or theistic evolutionists. There are three especially perplexing problems that accompany the notion of an entirely deathless creation before any human "fall." The first of these I will refer to as *the stasis dilemma*.

THE STASIS DILEMMA

The Creator God of Genesis clearly takes joy in untamed creaturely flourishing and procreation. All living creatures are commanded to be fruitful and multiply, and to spread out and fill the earth. In a spatially finite and deathless world, however, there could not be endless procreation. At a certain point in time, all births would cease and there would be no new species or changes in the precise number of creatures in existence. Sexual differentiation would then become superfluous and all reproductive organs in the animal kingdom would become vestigial oddities like tonsils or appendices. A "perfect" world in the sense some believers demand would, it seems, be a world without any young or old (or perhaps with young creatures that never grew to adulthood). It would be a world without nurturing, nesting or protecting of eggs, without rituals of courtship and mating, and without any new kinds of creatures ever emerging. It would be a creation without new creation. But would such a world really be "very good"? Or would it be, as John Haught writes, "dead on delivery"?[3] A flawlessly engineered world without growth, without new birth, without change and without death might be "perfect"—like a finely calibrated watch. But "perfect" in this sense might not be good at all for God's creative purposes.

The very language of flawless "design" employed by creationists, with its endless analogies to watches and other impressive human devices, often seems to be a theology inflected through an unwholesomely modernist grammar of nature as *mechanism*. Alternatively, we might posit a world in which creaturely flourishing *would* continue without end through utterly fantastic divine interventions we can in no way fathom (such as teleportation of animals to other planets or expansion of the planet to infinite size). But this would mean, in effect, a surrealist world with no conceivable relation to the world we actually live in. The idea of a deathless—and so, ultimately, *birthless*—creation might ironically be the ultimate expression of gnostic contempt for the Hebrew tradition and for God's actual creation in all of its complex, untamed, finite and often messy earthiness.

The stasis problem is real, though somewhat less severe, as far as the idea of a deathless humanity is concerned. The fact that Adam and Eve alone among the animals are made "in the image of God" and that God directly breathes into them alone "the breath of life" may include the idea that they uniquely participate in, or conditionally receive, something of God's own immortality. Humans are not bound to the earth in the same way the other animals are. They are (as the devil Screwtape reminds Wormwood in C. S. Lewis's *Screwtape Letters*) intermediate or "amphibious" creatures, at once animals and exalted spiritual beings.[4] But the rest of the animal kingdom, as far as we know, is purely earthy. This is their only home, and they do not fully share the moral and spiritual faculties that humans possess. This means that animal predation *by definition* is not "sinful" in the way human violence is sinful, even if nature has been marred or distorted or thrown into disarray in some sense by human sin; a lion devouring an impala (or Cape buffalo in my childhood memories from Mana Pools) is not morally culpable of interspecies murder—it is simply doing what lions do, and there is presumably no hell for unrepentant lions or heaven for their hapless prey (although some Christians have speculated that there may be a heaven for at least some animals).

Once we make a number of vital distinctions—between human nature and animal nature; between mortality and predation; between generous, self-emptying death and competitive, self-enhancing killing; between evolution with predation and evolution without predation; between evolution with a *telos* and evolution as a result of "blind algorithms" and strictly materialistic laws; between creation by divine fiat at every step and creation as a divinely appointed process that opens spaces for undetermined and potentially risky "secondary causes"—it therefore makes perfect theological sense to think of God's creation as mortal, free, ongoing and *very good* from the start. There is nothing in Genesis or the rest of the Hebrew Bible to contradict this idea and much to commend it.

Even passages in the New Testament such as Romans 5 that speak of death entering the world through one man are, on closer examination, exclusively focused on humanity.

> Wherefore, as by one man sin entered into the world, and death by sin; and so death passed upon all men, for that all have sinned: . . . Therefore as by the offence of one judgment came upon *all men* to condemnation; even so by the righteousness of one the free gift came upon *all men* unto justification of life. (Rom 5:12, 18)

If some readers are unhappy with restricting the meaning of Romans 5 to humankind alone, the question we must ask them is, have they considered the theological implications of expanding the meaning of the text beyond what it plainly says? Let us imagine that a well-meaning medieval scribe had inserted the following glosses on the text: By one man "death passed upon all men and animals, for that all men and animals have sinned: . . . by the righteousness of one the free gift came upon all men and animals to justification of life." Or what if the text read: "Through one man death passed upon all men and animals, for that all men have sinned and innocent animals were thus cursed by God . . . through one act of righteousness the free gift came unto all men but not to animals to justification of life." Would this tidily solve the theodicy problem of animal suffering and predation—or in fact render it incomprehensible?

According to Genesis, Adam and Eve were not created deathless or immortal in their physical persons. They were created as fully mortal beings from the start. Within the narrative universe of Hebrew Scripture, the only basis for humans avoiding death was their ongoing access to the fruit of the tree of life in the Garden of Eden, from which they were barred by the cherubim (fierce winged beasts in Near Eastern mythology) after their rebellion. This resulted in their returning, quite naturally, to the dust from which they (along with every other creature in the Garden) originally came. Humanity alone, a plain reading of the text strongly suggests, was granted special access to a provisional and renewable source of life. They alone then forfeited this access through the exercise of their free wills, falling downward (if not backward) into the realm of mortal animal existence.

The Deceiver God Dilemma

Another, more serious dilemma for literalistic hermeneutics is what we might call *the deceiver God dilemma*. Most young earth or young life creationists I have spoken with, when pressed about the weight of empirical evidence, concede that their models cannot explain the physical data from biology and geology in any kind of satisfying way, and that they would never have arrived at their views were it not for the fact that they begin with a very particular set of assumptions about how the biblical text must be read. They did not come to their scientific conclusions independently of their interpretations of Scripture and then discover that the two serendipitously coincided. Rather, they have "corrected" the lens through which they interpret all empirical evidence—the correct "lens" amounting to a dogmatic understanding of what Genesis "scientifically" means. We must uphold the inerrant scientific/historical truth of Genesis, we are told, because otherwise God's Word would be unreliable, untrustworthy and deceptive. Why would God say he created the earth and everything in it in six literal days if he did no such thing?

But if the creation accounts in Genesis 1–2 are in fact archetypal narratives that were never meant to be read in this kind of thoroughly rationalistic yet imaginatively impoverished way, the trouble would not only be that we have often viewed the empirical data through the distorting lens of philosophical materialism. It would be that we have also read Scripture through the distorting lens of biblical literalism and modern philosophical foundationalism. The moment we allow that there are many different ways of narrating truth, including mythopoeic ones, we realize that the question of divine deception only exists as a live problem for unbending literalists themselves (and mythopoesis must be carefully distinguished from mere fable, parable or spiritual allegory).

Confronted by the fact that there are continuous tree ring records dating back more than ten thousand years, the Institute for Creation Research offers the following blithe answer: "Trees were likely created with tree-rings already in place."[5] The implication is clear: "scientific" creationism is impervious to scientific evidence of any kind. Through the eyes of faith, even negative evidence may be chalked up as confirming evidence of one's prior convictions. But the question for creationists who brush aside compelling scientific data with conversation-stopping replies like this (or assurances that with just a little more time and effort the literalist position will be triumphantly vindicated by new scientific discoveries) is: Why would a Creator who required us to deny or suppress so much of our reason, our observations and our senses be more worthy of our trust and love than a

God who, as the sustaining ground of all being, permitted or guided evolutionary processes that we can partially observe and comprehend with our God-given minds, and that we can now grasp in *theological* terms through the creation narratives of Genesis?

The goddess Athena, according to the Greeks, sprang fully formed from the head of Zeus. Not to be outdone, young earth creationists have conceived a Creator who pulls fully formed rabbits and people from out of the soil on day six of the creation. But while this would be an impressive feat, we must ask what sort of picture of God emerges from this vision. Is this not a strictly nominalist or voluntarist God whose ways may indeed be stupendous but whose creation—as an inscrutable performance of sheer will—must now also be seen as a kind of deception or sleight of hand? What should we make of a God who creates a universe, an earth, plants, animals and humans with the *appearance* but not the actuality of age? How many days old did Adam *appear* on the first day of his creation? Those who would say more than one are positing a basic incongruity between the reality disclosed in Scripture and the realities of the physical world that stare us—and that stared Eve—in the face. "If anything at all has been truly created, it must have been created with an appearance of at least some prior history," declared Henry Morris. "It would only 'look' old to one who rejects the very possibility of special creation."[6] Existence at a very fundamental level, in this way of thinking, cannot be believed or trusted. It is at bottom an artificial stage production. Reality has the ontological properties of unreality or surrealism and did so from the very start.

What the nominalist Creator seems to require of us is *not* belief in the superabundance of divine love that opens the possibility of the miraculous as a revelation of what is in fact most natural and most real in God's inbreaking kingdom—the kind of hyperreality that might even raise a man from the dead. Instead, what the nominalist deity demands of us is a performance of sheer will in turn: the will to believe; fideistic mental compliance to purely propositional assertions in the name of protecting the Bible's internal coherence; unquestioning acceptance of a creation that now contains the arbitrary and surd elements of an unbelievable magic show (thousand-year-old trees that are really one second old, day-old humans without any memories who nevertheless know how to speak to one another in a fully evolved human language, apparently downloaded directly from the mind of God like preinstalled software).

The Divine Curse Dilemma

The third and most severe theodicy problem with biblical fundamentalism or literalism on the doctrine of creation, however, I will refer to as *the divine curse dilemma*. Entirely apart from the scientific evidences for predation before the emergence of human beings, we are confronted all around us by the plain fact of animals killing in order to survive. The natural world is filled with creatures that are anatomically "designed"—in their internal organs, their instincts and practically every fiber of their physical structures—to exist by consuming other creatures. Some of these animals we would have to characterize as *irreducibly predatory* (think parasitic wasps, the parasites that live inside the parasitic wasps, the viruses that infect the parasites that live inside the parasitic wasps—and of course the rodents that devour all of the above in a single gulp). And yet these creatures play a vital role in the cycles of life and death, the great economy of nature in which nothing is ultimately wasted, purposeless or "selfish" insofar as all creatures must die and in dying make it possible for other creatures to live. How, then, did these predatory creatures—and indeed, the entire natural world as we know it right down to the cellular level and basic metabolism—arise? One can imagine three possible replies (or a combination of them) to this riddle that would conform to highly literalistic interpretations of Genesis that assume all death in nature to have resulted directly from Adam's fall, but none are theologically coherent and each raises baffling questions about God's justice and character. I call attention to these problems not in order to criticize the faith of others but to challenge the pretentions to absolute certainty and authority that some Christians evince in discussions over creation and evolution. The plain fact is that the problem of theodicy, in light of animal predation and animal suffering, cannot be resolved simply by appealing to a literal creation week in the recent past or to Adam's decision to eat the forbidden fruit.

Possibility one. After humanity's rebellion, God gave the animal world over to natural laws of competitive rivalry, self-interest, chance and death, so that over time the instinctual behaviors, internal organs and physical structures of all creatures *evolved*, turning them into the predators we see today. The theological trouble with this answer is that the Creator is conceived of as a divinity who consigns countless morally innocent creatures from a state of natural bliss to one of suffering and death without any ability to comprehend the meaning of their transformation and for no redemptive purpose as far as the animals themselves are concerned. ("The deeper minds of all ages have had pity for animals," Nietzsche wrote, "because they suffer from life and have not the power to turn the sting of the suffering against themselves, and understand their being metaphysically."[7]) God therefore remains no less implicated in the harsh facts of nature than in an evolutionary paradigm, and indeed it would seem even more so.

For some theistic evolutionists, animal suffering and predation should be seen as "the birth pangs of creation" in a universe that includes not only free wills but also free processes within a divinely ordered but not rigidly deterministic framework (an idea I will return to). For biblical literalists who take the path of postlapsarian naturalistic evolution, however, the same realities must now be seen as the result of God's decision to *abandon* the animal world to the ravages of sin and contagion of human violence, even though the animal kingdom was in no sense morally responsible for human moral choice. But why, we must ask, should Adam's disobedience have required placid, plant-eating sea creatures to be rapidly transformed into great white sharks and killer whales plying the oceans in search of seals to be devoured in dark places where no human eye ever falls (not to mention velociraptors and tyrannosauruses)? What kind of just and loving God would not merely permit but positively *require* this to happen?

Further, under this naturalistic scenario the very evolutionary mechanism that literalists have long sought to dispense with through "scientific" creationism has ironically been reintroduced as a potent, creative force capable of generating entirely new species and stunning new adaptations in a remarkably short time—practically instantaneously, it would seem. If it were *natural* laws that produced these changes in the animal kingdom, the process should be open to investigation by strictly natural science. Where, though, is the scientific evidence for this massive and sudden postlapsarian *evolution* throughout all of nature?

In the face of such insurmountable scientific difficulties, intelligent design theorist William Dembski declares that for theological reasons we must continue to attribute all natural evil to Adam and Eve's historical fall but for scientific reasons now do so *retroactively*. In his foreknowledge, Dembski suggests, God created animal predation many millions of years before the creation of humanity in order to somehow (Dembski does not explain just how) contain as well as punish the effects of human sin, not unlike firefighters setting "backfires" in anticipation of a coming blaze. "God brings about natural evil," Dembski writes, "to free us from the more insidious evil in our hearts."[8] God is not bound by time or by laws of linear causality. Adam and Eve thus remain morally responsible for all evil in the world, including death in nature *before any humans existed*. But while Dembski's ideas might help to ease some of the scientific challenges faced by strict biblical literalism I fail to see how they ease literalism's theodicy dilemma. In the end, he simply repeats in a somewhat unconventional form what is perhaps the most common literalist reply to the question of the origins of animal suffering and predation: God imposed a "curse" upon all of nature to punish rebellious humans for their disobedience. The time has come to examine this idea more closely.

Possibility two. After the fall of humanity, God supernaturally modified or "cursed" (i.e., *zapped*) the entire animal kingdom as a punishment for Adam's sin, creating the great predators we see today fully equipped with sharp incisors, talons and claws, and digestive tracts capable of processing only meat. If God could create the universe *ex nihilo* in the beginning, he could surely miraculously transform the creation after the fall if he desired. Such is the position, for example, of the seminary faculty of Andrews University in Berrien Springs, Michigan. In their 2010 "Statement on the Biblical Doctrine of Creation," they declare that, "The consequences of the Fall were severe, not only for Adam and Eve but for the entire world over which God had given them dominion. A curse was pronounced upon the animal and plant world and upon the ground."[9] Other conservative evangelicals have used still more vigorous language. According to Robert Hughes and J. Carl Laney in their *Tyndale Concise Bible Commentary*, "In Genesis 3 God cursed everything he had made and damned it to an eternity in hell."[10] The reason for the divine "curse" upon all animals, creationist Nicholas Miller meanwhile explains in more temperate but still troubling terms, was a kind of moral pedagogy for humans. "The thorn and the thistle, nature red in tooth and claw, show the outworking of the philosophy of egoism and selfishness and are part of man's education, making him aware of the importance of the principles of God."[11]

This explanation for animal suffering neatly avoids all scientific difficulties by instructing us to simply have faith in God's power to do anything according to his sovereign will—including miraculously turning docile animals into ferocious predators for the edification of disobedient humans in the effects of "egoism and selfishness." Under this schema, unfortunately, all the theological and moral hazards of possibility one above are retained and amplified. Puritan theological determinist Jonathan Edwards similarly suggested that we should read nature in radically anthropocentric terms as a kind of didactic, typological lesson book designed to educate fallen humanity in God's sovereignty and the consequences of their depravity.[12] But if the Creator "cursed" all animals "as part of man's education" in "the principles of God," what exactly are the principles we should take from the lesson? Instead of natural processes and principles of freedom and finitude helping to explain the finely tuned balance between life and death that we observe in all of biological existence, God himself must now be seen as the active designer of every adaptation to shred, tear, dismember and digest an organism's prey. But what kind of Creator would punish Adam and Eve's rebellion—whether retroactively or proximately—by bending the rest of his creation from a state of perfect peace into so many malign forms, supernaturally summoning into existence the snake's venom and the jaguar's teeth and commanding innocent creatures to begin devouring one another for the moral instruction or chastisement of humans? What would we think of a parent who decided that the best way to educate their child in the combustibility of fire was to place the family cat on the stove? The child might learn something about fire, to be sure. But what would they learn about their parent?

If the response of the biblical literalist to these quandaries is that we must simply trust and obey, we are compelled to ask them what exactly they would have us put our trust in. As was pointed out in chapter one, the notion of God actively cursing all animals may be a longstanding part of the Christian tradition. But on close reading we must confess that it is actually not included in Genesis or any other part of Scripture (unless by questionable interpretation of scant and ambiguous verses). Not even Adam and Eve are "cursed" in Genesis. Why would God then curse innocent animals to punish morally guilty humans? The only living being that is "cursed" in the Genesis narrative is the serpent—that is, the one who bears direct moral culpability for the human fall and who, according to later Jewish and Christian tradition, was not an animal in any case but a disguised form of the "accuser" angel, *ha-satan*.

Possibility three. God did not miraculously create animal predation after the fall, nor did predation arise over time as a result of natural, evolutionary laws. Rather, it is the result of demonic biological experimentation or gene manipulation, perhaps even with the connivance of depraved, antediluvian evil scientists working in hyperadvanced genetics laboratories, the evidence of which was unfortunately (or conveniently) destroyed in the flood—an actual theory for the origins of the dinosaurs entertained by some I know![13] Yet what do claims such as these of diabolical postlapsarian countercreation really say about God and about our universe? Are we still at this point within the fold of orthodox Christian and Jewish faith? Or have we entered the realm of gnostic speculations (if not science fiction) that give far greater creative power to forces of satanic agency than Scripture ever does? This way of resolving the immediate problem of animal suffering might offer some short-term relief to some creationists. But does it offer any kind of constructive long-term paradigm for doing rigorous scientific research? The concept of demonic or satanic supernatural reengineering of animals for carnivorousness in the recent past is no part of the Genesis narrative or of the rest of the Hebrew Bible. It also seems to be an intellectual cul-de-sac for anyone trying to conduct serious scientific research on questions of origins. "Intelligent design" theory and "scientific" creationism have failed to convince the overwhelming majority of scientists that they are progressive research programs. Will science as well as theology be helped or hurt by now invoking intelligent *malevolent* design? And by what criterion will creationists distinguish divine from diabolical design in nature—particularly when aspects of ferocity and beauty are so completely intertwined as to be inseparable parts of the same reality?

11

A Midrash

C. S. Lewis's Cosmic Conflict Theodicy Revisited

We must consider, though, a more sophisticated form of the argument from diabolical agency or "cosmic conflict" to help account for animal suffering. In his 1940 book *The Problem of Pain,* C. S. Lewis offered an approach to the question that includes such an idea, one that I think still contains valuable insights, although it is an approach that I also want to challenge in certain ways.

The first step in Lewis's theodicy in the face of animal predation/suffering is to make clear that anything we say on the subject will be highly speculative since the inner lives of animals remain a great mystery to us. We must carefully distinguish, Lewis points out, between sentience and consciousness. A sentient creature might pass through a series of discrete sensory states N, A, P and I. But a *conscious* creature is able to in some sense stand outside of its own sensations and connect them together as an experience: P-A-I-N. A conscious creature, in other words, has selfhood. It has subjectivity. It is quite possible, then, that there is no pain or suffering—and so no problem of theodicy—in lower-order creatures that are sentient but not "conscious." "It may be we who have invented the 'sufferers' by the 'pathetic fallacy' of reading into the beasts a self for which there is no real evidence."[1]

To this preliminary observation of Lewis's we might add another closely related one. Biblical literalists and Darwinians alike have burdened evolution with adjectives such as "cruel," "vicious" and "selfish." Yet these descriptions, we must see, are projections of human moral value onto nature that on closer examination might not be at all legitimate to make—at least not for predatory or evolutionary processes tout court. When an eagle catches a salmon out of a river to feed to its young, is it correct to describe the event as "vicious," "evil" and "selfish"? Or is it we who have invented not only the suffering of the fish by the pathetic fallacy, but also the "cruelty" of the eagle through the fallacy of reading into all forms of predation a kind of moral egoism for which there is no real evidence either? (It might seem a very strange thing to say, but one cannot help but also recall here that Christ, without being a "vicious" or "selfish" predator, not only ate fish but also delivered to his disciples a superabundant harvest of fish as a miraculous blessing and sign of his inbreaking kingdom. God in human flesh was sinless and he was also an omnivore.)

It is, nevertheless, very hard to imagine that all animals are merely sentient and not conscious (following Lewis's definitions of these terms). Naturalists have presented compelling evidence for animal emotions ranging from fear to grief to joy to shame to rage to compassion. We do not know at what stage in the ladder of animal development or by what processes sentient creatures become fully conscious beings. But we have good reasons to believe that primates and other higher-order mammals with highly developed neocortices such as elephants, dolphins and dogs have experiences of pain and suffering analogous to our own—and the possibility that lower-order creatures such as fish also experience real suffering cannot by any means be discounted. In a brilliantly reported and morally disturbing story on the Maine Lobster Festival for *Gourmet Magazine,* David Foster Wallace pointed out that even these insect-like crustaceans have complex nervous systems and react to being boiled alive in a way that clearly suggests they are in pain. Their behavior is so alarming that many cooks set clock timers and leave their kitchens until after the creatures are dead. At the end of all our intellectual abstractions about the difference between sentience and consciousness, Wallace writes, "There remain the facts of the frantically clanking lid, the pathetic clinging to the edge of the pot. Standing at the stove, it is hard to deny in any meaningful way that this is a living creature experiencing pain and wishing to avoid/escape the painful experience."[2]

In the absence of clear knowledge of the inner lives of other creatures, shouldn't we therefore follow a precautionary approach and assume until proven otherwise that the distinction between sentience and consciousness does not exist (or is one we simply cannot afford to indulge in the cases of most if not all animals)? We must guard against any view that, from the standpoint of a more enlightened age, might appear little better than the Cartesian doctrine that animals are complex machines. Descartes's follower Nicolas de Malebranche, it is said, was so impressed by the teachings of the new science that he viciously kicked a pregnant dog in the stomach to illustrate to his companions how lifelike a machine can be in its cries, and to impress upon them the fact that they should only be concerned with the "real" pain of human souls. In his memoirs published in 1738, Nicholas Fontaine described one laboratory that had embraced the Cartesian teaching that humans alone have minds and feelings worthy of our concern:

> They administered beatings to dogs with perfect indifference, and made fun of those who pitied the creatures as if they felt pain. They said the animals were clocks; that the cries they emitted when struck were only the noise of a little spring that had been touched, but that the whole body was without feeling. They nailed poor animals up on boards by their four paws to vivisect them and see the circulation of the blood which was a great subject of conversation.[3]

Although such beliefs were extreme even in the seventeenth and eighteenth centuries, they were nevertheless possible then as they no longer are today for a plain reason: Darwin's theory. Pre-Darwinian thinkers could still cling to the notion of an absolute discontinuity between humans and other animals despite their evident similarities and so avoid having to face the theological dilemma of animal suffering. But "once we see the other animals as our kin," James Rachels writes, "we have little choice but to see their condition as analogous to our own."[4] There is a disturbing (and I think not entirely accidental) historical relationship between Darwinism and projects of dehumanizing social engineering. But Darwin's theory also inspired many people to begin to think much more humanely about non-human animals.

In the Oxford don's defense, Lewis—who not only wrote books about talking animals but also fought against animal vivisection—stands within an important Christian subtradition that rejects anthropocentrism and declares the animals (following St. Basil the Great) to be "our brothers." He goes so far as to suggest in *The Abolition of Man* that there is deep impiety in doing harm even to trees.[5] The evolution of "nonconscious" creatures without complex brains, then, *might* be conceived as a process that does not involve either suffering or cruelty, which would take us some way toward answering the problem of theodicy in evolutionary perspective. But we still sense in the destructiveness of nature and in the suffering of animals even as lowly as fish and lobsters that the problem of evil has touched not only humanity but the animal world as well. God's good creation is groaning and in travail. It is *fallen* and it has been for as long as we can tell it has existed, ages before the appearance of human beings.

Lewis's way of dealing with this fact is neither to offer a general scientific theory nor a dogmatic reading of Genesis. Instead, he cautiously recalls, with what might be described as *faithful agnosticism* on some of the deepest riddles of the creation, certain Dominical, Pauline and Johannine references to the first creature to rebel against the Creator. The problem of evil, orthodox Christianity has long maintained, predates human existence. It is thus entirely possible to think of the animal world as having been "corrupted" or placed in "bondage" long before the first humans. "If this hypothesis is worth considering," Lewis writes,

> it is also worth considering whether man, at his first coming into the world, had not already a redemptive function to perform. . . . It may

> have been one of man's functions to restore peace to the animal world, and if he had not joined the enemy he might have succeeded in doing so to an extent now hardly imaginable.[6]

Lewis (perhaps wisely) does not pursue these speculations further. We might observe, however, that when Adam is told by God to "till and to keep" the Garden, the Hebrew word for "keep" is *shamar*, which means not only "to preserve" but also "to *guard*." Adam must protect the Eden sanctuary, it seems, from incursion and harm. Is part of Adam's task to "keep" a portion of the creation safe against perils that already exist in the larger world as figured by the serpent and to "subdue" these elements in due time?

We moderns (and postmoderns) of course instinctively balk at the notion of a personified malevolence or evil—an "enemy" or anti-Adam in the universe—as well as at the undeniably biblical language of spiritual warfare involving intelligences in God's created order other than human beings and animals. And there is a real danger in these ways of thinking of sliding into a sheer dualism, gnosticism or Manichaeism. Nevertheless, there is a clear sense throughout the New Testament that are we living in the time of a temporary dualism in which God has permitted parts of his creation—and not humans alone—the autonomy of radical freedom and even defiance, which God himself must now in some sense struggle against. ("If it offends less," Lewis wrote, "you may say that the 'life-force' is corrupted where I say that living creatures were corrupted by an evil angelic being. We mean the same thing: but I find it easier to believe in a myth of gods and demons than in one of hypostatized abstract nouns."[7]) Evil in biblical thinking is ultimately related to a spiritual being of great destructive will who possesses a backstory with narrative depth that includes powerful accusations concerning the justice and goodness of God's creation and God's sovereignty. At the same time, N. T. Wright insists, it would be wrong to think of the demonic as personal in the same sense as Jesus or the disciples. The evil personified in biblical language as *ha-satan* does not have the full dignity of personhood but is rather "sub-personal" or "quasi-personal," "the moral and spiritual equivalent of a black hole."[8]

How do these "mythical" ideas help to shed light on contemporary questions of faith, science and theodicy? Believers who take seriously the idea of the diabolical have perhaps been unduly influenced by the geography of John Milton's *Paradise Lost*, which begins with Satan and his legions languishing in Pandemonium, the capital of hell. It is only after an arduous journey across the universe, in Milton's telling, that the devil discovers and infiltrates newly created Planet Earth, wreaking havoc on God's good work through Adam and Eve's rebellion. The book of Revelation, by contrast, traces a very different arc in Satan's "fall like lightning" (in the words of Christ in the Gospel of Luke). Unlike in Milton's poetics in which hell is situated a vast distance from earth, requiring that the devil (anti)heroically traverse the galaxies in search of unspoiled territory, in the mythos of John's Apocalypse Satan and his legions are cast from heaven to one place alone:

> And there was war in heaven: Michael and his angels fought against the dragon; and the dragon fought and his angels, And prevailed not; neither was their place found any more in heaven. And the great dragon was cast out, that old serpent, called the Devil, and Satan, which deceiveth the whole world: *he was cast out into the earth, and his angels were cast out with him.* (Rev 12:7-9, emphasis mine)

These mysterious verses suggest a widely overlooked fact: the earth itself in a certain sense is the only "hell" that has ever existed. We find ourselves situated both spatially and temporally in the ravaged battlefield, home and dominion of "the prince of the power of the air" (Eph 2:2).

Such a reading would clearly give new meaning to the apostle Paul's insistence that Christ is the Second Adam, the one the creation waits eagerly for, the long-foretold redeemer who will at last "set free from its slavery to corruption" the "whole creation," which "groans and suffers the pains of childbirth until now," the "firstborn of all creation" whose *euangelion* must be proclaimed "in all creation under heaven" (Rom 8:19-22; Col 1:15, 23 NASB).

According to the book of Revelation, Christ is the "beginning of the creation of God" and the lamb "slain from the foundation of the world" (Rev 3:14; 13:8). If we take this language not only of human but of cosmic redemption seriously, we will see that the gospel is not only good news to people—it is good news for creation in its entirety, including suffering and stupefied animals, subjected to chaos, cruelty and death not by their own sinfulness, nor by Adam's disobedience, nor again by God's design, but potentially before the arrival of humanity in a universe of unequal but mysteriously conflicting spiritual realities.

The difference between this way of thinking and the notion of supernatural demonical reengineering of animals in the recent past that I referred to earlier may at first seem slight, but I believe is significant. Human as well as natural history now appears as the stage for a drama that has involved opposing principles of freedom and sovereignty for vastly longer than we may have first imagined. This drama still awaits its final act, which promises nothing less than a great transformation of all of creaturely existence as the material and temporal is at last delivered from its captivity to the "principalities" and "powers" and is caught up in God's eternity. It is not a matter of evidential proof, logical necessity or scientific knowledge (or *gnosis*) but of Christian faith that we find ourselves in a shadow land that is a dim reflection of God's original creative intent and final redemptive purpose. It is entirely within orthodox Christian thought to think of these shadows as having been cast long prior to the appearance of human beings. Ours is an earth and possibly a universe that has in some sense been "inverted," distanced, even violently separated from the source of all life, so that while nature still bears a fractured witness to its origins in the divine love it is also subject to the destroyer "god" of this world (2 Cor 4:4) whose unyielding laws might produce spectacular flourishes for a time but that ultimately can tend only ever downward into the nothingness that is complete deprivation of the good.

There is nothing in the Lewisian cosmic conflict scenario, we might note, to contradict the evidences from science of a very old earth, very old life on earth, a great deal of common ancestry among organisms including humans and other creatures, and animal predation before the appearance of human beings. But neither is there anything in this telling to rule out the literalist claim that humanity *qua* humanity began with a single couple (a view that Lewis seems to have held). Material structure, many literalists appear to have forgotten, is not what the *imago Dei* refers to. Hence, John Stott points out, acceptance of Adam and Eve "as historical is not incompatible with [the] belief that several forms of pre-Adamic 'hominid' may have existed for thousands of years previously":

> It is conceivable that God created Adam out of one of them. You may call them *homo erectus*. I think you may even call some of them *homo sapiens*, for these are arbitrary scientific names. But Adam was the first *homo divinus*, if I may coin a phrase, the first man to whom may be given the Biblical designation 'made in the image of God.' Precisely what the divine likeness was, which was stamped upon him, we do not know, for Scripture nowhere tells us. But Scripture seems to suggest that it includes rational, moral, social, and spiritual faculties which make man unlike all other creatures and like God the creator, and on account of which he was given 'dominion' over the lower creation.[9]

Although Lewis's cosmic conflict theodicy will undoubtedly prove far too literal-minded for many believers and not nearly literalistic enough for others, it is valuable, it seems to me, as a way of helping to mediate at least some of the disagreements between those literalists who are still committed to honest dialogue and those who also read the book of Genesis as an authoritative account of God's creative activity but who do not feel bound to notions of scientific/historical infallibility or chronology to explain what Genesis literally means. It is perhaps helpful to think of Lewis's writing as recovering a vital form of biblical interpretation well known to the Jewish faith as a *midrash*.

An interpretation that is "midrashic" is not concerned with convincing readers that this was the way things *must* have happened so much as it is with telling stories about stories as a way of opening new ways of thinking about the *silences* of the text, about its present absences. The advantages of Lewis's in some ways highly literalistic but at the same time nondogmatic speculations include the following: (1) he emphasizes competing principles of freedom rather than postfall miraculous refashioning of matter (whether demonic or divine) to account for the physical universe we now see before us—an approach that seems to this reader to offer greater theological and moral (not to mention scientific) coherence; (2) he offers a nondefensive and open approach to what scientists have to tell us about the evidences of the "book" of nature, without giving away to philosophical naturalists the conceit that they possess the full story; and (3) he exhibits a humble recognition that there is much we simply do not know or understand on both the scientific *and* the biblical sides of the problem.

Biblical literalists and Darwinian materialists alike share a common zeal for constructing airtight systems of knowledge on foundationalist epistemological assumptions that leave little room for ambiguity, perplexity, mystery or *poetry* in our worldviews. The Lewisian approach, in which Genesis is read as a "true myth" (or "creation saga" in Barthian terms) that demands not scientific proofs so much as a reframing of imaginative horizons, has the great advantage of relieving us of an oppressive burden: the burden of knowing. We have a great deal of evidence for evolutionary processes at work in nature, although the causal mechanisms of this evolution and their reach are far less clear. We do not fully know what Genesis means or how to fully resolve the problem of animal suffering, both in the past and the present, in the light of the biblical creation accounts and evidences from biology and geology. We also do not know all that God has and has not permitted to unfold from the beginning of time in his untamed universe.

Thankfully, we do not need to have complete answers to these questions in order to have faith or to continue to investigate scientific problems with scientific integrity, following the empirical evidence wherever it may lead and not predetermining where it must lead based on what we think we already know. The exhausted but interminable conflict between faith and science, Lewis demonstrates, arises only for those believers who have conceived of Genesis as a *systema naturae* that must now be used to somehow coordinate all scientific knowledge related to questions of origins on the one hand, and those philosophical materialists who have embraced the scientific method as a kind of new revelation that can be used to map all of reality on the other.

12

GOD OF THE WHIRLWIND

Animal Ferocity in the Book of Job

STILL, I MUST CONFESS NO SMALL UNEASE at Lewis's speculative cosmic conflict theodicy (which has been perhaps most vigorously defended in recent times by Gregory Boyd[1]). For one thing, while we should not be quick to dismiss the idea of nonhuman rebellious powers sowing destruction in the cosmos over long periods of evolutionary history, as John Polkinghorne points out, the theodicy problem remains very much unresolved since there is no clear answer to the question of how these dark powers originated and why God should have permitted them to wreak such havoc for so long.[2] In attributing most if not all animal suffering to diabolical agency I also cannot help but wonder if Lewis has given the devil more than his due. The trouble is, when we search elsewhere in both the Hebrew Bible and New Testament we find little evidence that the biblical writers conceived of all animal suffering as a marker of "sin" or demonic corruption of the material forms of creation. This may be because they were simply not sensitized to animal suffering the way we are today. Or it may be that they thought carefully about the question and arrived at a very different conclusion. According to the Gospel of John, God created not only those parts of the world that measure up to our moral approval or understanding. "*All* things were made by him; and without him was not any thing made that was made" (Jn 1:3). Hence, Kentucky farmer-philosopher Wendell Berry concludes in what I think is a serious reading of the evidences from both nature and Scripture, "we must credit God with the making of biting and stinging insects, poisonous serpents, weeds, poisonous weeds, dangerous beasts, and disease-causing organisms." "That we may disapprove of these things," Berry continues, "does not mean that God is in error or that He ceded some of the work of Creation to Satan; it means that we are deficient in wholeness, harmony, and understanding—that is, we are 'fallen.'"[3]

Any answer we give to the problem of animal predation or animal ferocity within God's good creation must wrestle with the most extended commentary on the creation process/event in the biblical canon outside of Genesis itself. There is another book in the Hebrew Bible that talks about the creation in great detail and that some scholars believe was written even before the book of Genesis, which would make it the first creation account in Scripture, even though it has not often been thought of in these terms. I am speaking of the final four chapters of the book of Job. God's reply to Job from out of the whirlwind is Jewish Scripture's clearest answer to the problem not only of human but also of animal suffering. And it is a serious challenge to creationists and Darwinian evolutionists alike.

At the heart of Job's protest against God beginning in Job 3 we find the distressing idea that it would have been better if God had not created at all. Rather than accept a suffering creation and his own sufferings within it as "very good," Job calls for the creation in its entirety to be undone. "Let the stars of the twilight thereof be dark; let it look for light, but have none; neither let it see the dawning of the day," he declares (Job 3:9)—an exact reversal of God's "Let there be light." God never cursed all of the creation. Job now does. The challenge posed by Job is not simply the problem of distributive justice: Why do the innocent suffer? It is the problem of nihilism: Why is it better that there should be a suffering creation rather than no creation at all? God might have made an earth in which there was creaturely flourishing and happiness without any possibility of suffering or death. How, then, can God proclaim the world he actually created to be "very good"? Job does not deny God's existence on skeptical and evidential grounds but rather declares that the costs of creaturely existence are too high. He accepts the *reality* of God's existence but, like Ivan Karamazov in Fyodor Dostoevsky's novel *The Brothers Karamazov,* he respectfully "returns the ticket."[4]

Strictly speaking, Job's case against God's created order—not simply the problem of suffering or distributive justice but the challenge of nihilism—is unanswerable in any purely rationalistic or moralistic terms. If one comes to the book of Job expecting or demanding such an answer, God's words from out of the whirlwind will appear as little more than the tirade of a bullying tyrant. Instead of abstract concepts, however, God's answer to Job's nihilism—to Job's will to nothingness—is nothing other than the creation itself in all of its stupendous, intricate, frightening, free and often incomprehensible forms. In one sense this is not an answer to the problem of suffering at all—certainly not Job's *personal* sufferings. But in another it is the only answer possible. The creation, with its suffering and death included, is very good because it is *God's* creation. There is an order of existence that is mysteriously, terribly and wondrously God's own—an order that cannot be anthropomorphically cut down to the measures of human scientific reasoning *or* human moral fastidiousness and systematic theology. Job is right to cry out in protest against his own sufferings; yet in turning his personal experience of suffering into an indictment against the creation in its entirety—against the injustice of *existence*—he goes too far. This is why, it seems to me, God both praises and rebukes Job when at last he speaks from out of the whirlwind.

"Where wast thou when I laid the foundations of the earth?" the Creator of Hebrew Scripture asks the "scientific" creationist and the Darwinian evolutionist alike (Job 38:4). "Have the gates of death been opened unto thee? or hast thou seen the doors of the shadow of death? Hast thou perceived the breadth of the earth? declare if thou knowest it all" (Job 38:17-18). The God who speaks from out of the whirlwind "caused the dayspring to know his place; That it might take hold of the ends of the earth, that the wicked might be shaken out of it" (Job 38:12). Yet there is no hint of wickedness or "natural evil" in the wildness and even ferocity of the animal kingdom. These aspects of his creation God seemingly delights in.

The Lord is the one who has carved "a way for the lightning of thunder" (Job 38:25). He causes "it to rain on the earth, where no man is; on the wilderness, wherein there is no man; To satisfy the desolate and waste ground" (Job 38:26-27). The Creator provides meat to the ravens, which are both scavengers and predators (Job 38:41).[i] He is the one who helps wild donkeys to escape their masters and gives them "the wilderness, and the barren land" for a home (Job 39:6). The ostrich "is hardened against her young ones" and does not tend to her eggs because God has not "imparted to her understanding" (Job 39:16-17). The Lord commands the eagle to "make her nest on high" from where "she seeketh the prey" so that "Her young ones also suck up blood: and where the slain are, there is she" (Job 39:27, 29-30). We see God's grandeur and wisdom in "the treasures of the snow" and "the treasures of the hail," in fearless warhorses whose necks are "clothed with thunder," and in the Behemoth and the Leviathan (Job 38:22; 39:19). The New English Bible renders God's lavish praise of the Behemoth (the grand finale in his reply to Job's questioning) thus:

> Consider the chief of the beasts, the crocodile, who devours cattle as if they were grass: what strength is in his loins! What power in the muscles of his belly! . . . He is the chief of God's works, made to be a tyrant over his peers; for he takes the cattle of the hills for his prey and in his jaws he crunches all wild beasts. . . . I will not pass over in silence his limbs, his prowess and the grace of his proportions. . . . He has no equal on earth; for he is made quite without fear. He looks down on all creatures, even the highest; he is king over all proud beasts. (Job 40:15-34)

---

[i]We might detect theological resonances to this passage in Job in Christ's declaration in the Gospel of Luke: "Consider the ravens: for they neither sow nor reap; which neither have storehouse nor barn; and God feedeth them" (Lk 12:24).

The psalmist declares that God made humanity "to have dominion over the works of thy hands; thou hast put all things under his feet" (Ps 8:6). But in Job, the Lord of creation radically humbles all human pretensions to mastery and control over nature by declaring that the *crocodile,* not the human, is king over all the beasts. What is more, the Creator takes full responsibility for animal predation, and there is no hint that it is anything other than *very good.*

Earlier in the poem, Eliphaz the Temanite—the spokesman for a false theology claiming to defend God's character and justify the ways of God to man—describes God as the one who breaks the teeth of young lions, scattering their young so that they perish "for lack of prey" (Job 4:11). But when God speaks for himself he declares that he is the one who *provides* prey for the lions, guiding them in the hunt (Job 38:39). The fact that all of these statements are in the form of poetry rather than of systematic theology should not prevent readers from seeing them as developing a vitally important account of origins and theology of creation. The God of Job is not a God who glories in defanged lions, which is to say, unlions. Isaiah 11:1-9, by contrast, envisions a future peaceable kingdom in which "the earth shall be full of the knowledge of the Lord" and "the wolf also shall dwell with the lamb, and the leopard shall lie down with the kid . . . and the lion shall eat straw like the ox." But the Isaiah passage, unlike Job 38–42, contains no parallel language, allusions or references to the Genesis creation. Its orientation is strictly apocalyptic, anticipating a final transformation of the creation without providing any commentary on its origins.

The God of the whirlwind—the God who takes responsibility for all of the creation in all of its strange, bewildering, endlessly innovative and untamed processes—may leave us perplexed and dismayed. But lest we question the justice or goodness of God's ways in creating the eagle, the lion and the great sea monsters, we should ponder the verse that follows closely after the poem's vivid description of eagles feeding their young the blood of other animals. "Will the faultfinder contend with the Almighty?" God demands of Job (Job 40:2 NASB). It is a question we must continue to ask ourselves today. Classical rabbinical hermeneutics, especially during the period of the Babylonian exile, included a method known as *targum* that involved imaginatively retelling and expanding upon the ancient biblical texts in more contemporary idioms.[5] Without calling it this by name, William Brown offers the following *targum* on the final chapters of the book of Job:

> Job . . . fasten your seatbelt and let us travel, you and I, into the dark, cold depths of another world, free from the propellers and harpoons of the surface, free from the "toil under the sun." . . .
>
> Behold the enigmatic *Grimpoteuthis*. Humans call it the Dumbo Octopus, though they are quite confounded about what it does in the deep. It simply rests on the bottom, wrapped in its mantle. Job, do you know what it does sitting so still and quietly in the dark? Answer me, Job, for surely you know! No? All right, then, I'll let you in on a secret: It's meditating on the Torah! . . .
>
> But my favorite creature of the deep is the one that humans disparagingly call *Vampyroteuthis infernalis*, "the vampire squid from hell," so named because it so repulsed its first discoverers. But it is my mascot of the deep: half-squid and half-octopus, dating back to 200 million years ago. Oh yes, you were born before then, weren't you Job? This creature can do something no other complex creature can: it can dwell quite happily in the oxygen-depleted layer of the ocean because of its special respiratory blood pigment. Being the slowest cephalopod of the sea doesn't hurt either.[6]

The vision of creation in the book of Job, Kathryn Schifferdecker suggests, is unique in the Hebrew Bible in its "radical non-anthropocentricity."[7] When Darwinian theorists and creationists alike declare that the natural world revealed by modern science is too wild, too finite and too ferocious to be God's very good creation, is it possible that they are simply repeating the nihilism of Job in his curse on the created world on the one hand and the false theodicy of Job's friends on the other? Is it possible that they are both, ironically, faultfinding and contending with the Almighty? As Robert Alter writes, "Job has been led to see the multifarious character of God's vast creation, its unfathomable fusion of beauty and cruelty, and through this he has come to understand the incommensurability between human notions of right and wrong and the structure of reality."[8] The theology of creation we find in Job does not exclude forces of chaos that are dangerous to human beings as well as other animals. But submission to God, Schifferdecker concludes, means learning to "live in the untamed, dangerous, but stunningly beautiful world that is God's creation."[9]

13

Creation & Kenosis

Evolution and Christ's Self-Emptying Way of the Cross

And yet there remains a deep scandal in death and suffering in nature that we must not allow the inspired poetics of the book of Job to cause us to forget or to become comfortably adjusted to. There are things under heaven and in earth that we should not be at peace with, and the jaws of the Behemoth, I would submit, are one. I have seen crocodiles on the riverbanks of Masai Mara in Kenya, near the end of the wildebeest migrations, their bellies distended from feasting. It is said they continue to kill even after they are engorged, without any interest in eating their prey. There is a turn in the Mara River where the wildebeest herds often cross and where by early November desiccated carcasses litter the banks, to be picked over by Marabou storks, maggots and flies. One can smell this open graveyard and hear the din of the birds from some distance. Some of the corpses lie partially submerged, their horns protruding from the fetid brown water where they were trampled in the stampede or ravaged by the massive reptiles. Calves sometimes manage to cross the river only to find themselves trapped by its steep banks. They drown in exhaustion amid the bellowing of thousands of their kind preparing to plunge after them into the murky water. These are the realities we must add our "Amen" to if we grant the God of the whirlwind who glories in the Behemoth and the Leviathan the final word. But on the banks of the Mara River, one's conscience might very well balk.

Perhaps Slavoj Žižek has discerned a vital truth in his provocative rereading of the book of Job not as a story of divine power over the creation but instead, in a certain sense, of divine impotence within it. God "solves the riddle by supplanting it with an even more radical riddle, by redoubling the riddle," Žižek declares, "he himself comes to share Job's astonishment at the chaotic madness of the created universe."[1] God's answer from out of the whirlwind amounts not to a negation but an intensification of Job's protest. What God is in effect saying, Žižek proposes, is that he too has no rational answer for the creation, that he is suffering along with Job. If God sounds slightly irritable it's because he's really just trying to hold it all together! But Žižek (a self-described atheistic materialist) goes still further, pressing the final chapters of Job in the direction of a radically christocentric interpretation that sees Job's silence at the end of the book as being filled with the pathos of one survivor bearing prophetic witness to the sufferings of another:

> What Job suddenly understood was that it was not him, but God himself who was in effect on trial in Job's calamities, and he failed the test miserably. Even more pointedly, I am tempted to risk a radical anachronistic reading: Job foresaw God's own future suffering—"Today it's me, tomorrow it will be your own son, and there will be no one to intervene for him. What you see in me now is the prefiguration of your own Passion!"[2]

Whether or not we accept this interpretation, we must confess that there is nothing in the reading of Job I offered earlier that a devout Jew or Muslim could not affirm. But Christianity—the faith whose central event is the brutal execution of the God-forsaken God on a Roman cross—greatly complicates and deepens our understanding of the divine response to suffering, whether of humans or of animals. It also denies us any stoical pact with the cruelties of death as divinely fated necessities of life. Death is the final enemy.

The most constructive approach to the theodicy dilemma of animal suffering, it seems to me, is the one taken by those theologians who have come to read Genesis and the evidences of natural science through a theological paradigm centered upon Christ's *kenosis* or self-emptying on the cross, and the ancient patristic understanding of *theosis*—the view that God's purposes in creating included his desire, from the beginning, for the divinization of humankind through the hominization of Christ. The creation was never a static golden age but always an unfolding story with an eschatological horizon. And the divine love has always willed that the journey of creation and pilgrimage of humanity should end in our final adoption as coheirs of God's kingdom and "partakers of the divine nature." The destiny of humankind is not simply a recapitulation or recurrence, paradise lost, paradise restored. Rather, the end is greater than the beginning—and was always meant to be so through the mystery of the incarnation.

One striking implication of biblical literalism is that Genesis tells us everything we need to know about God's way of creating without any reference whatsoever to the Christ of the New Testament. God's stupendous might, God's total control, God's complete domination of the creation by sheer fiat—such are the divine attributes that most impress the literalist and fundamentalist religious imaginations when they open the book of Genesis.[i] Yet there is in fact nothing intrinsically *christological* in these "plain" reading approaches to Genesis 1 or in the sorts of "scientific" and lexical arguments most often used to advance them. One can be a strict literalist on Genesis without possessing a trinitarian understanding of the divine nature and without any reference to the God who walked among us, whose power and

[i]Tellingly, the same literalists who vehemently oppose theistic evolution on theodicy grounds are no less adamant when we arrive at the book of Joshua that we must accept without question YHWH's commanding the Israelites to commit genocide of the inhabitants of Canaan—women, children, the elderly and animals. While there may be significant differences between the two problems, this seeming volte-face in moral concern for the suffering of the innocent (what did Canaanite cattle have to do with the sins of their masters?) suggests that it is an essentially divine command ethic rather than deep anguish at the realities of human or animal suffering that is driving literalist interpretations in both cases.

glory are paradoxically revealed in his weakness and agony. Literalist logic is strictly linear, requiring no rereading of what comes first in the light of what comes after. Perfect creation (C), we are told, is followed by fall (F) is followed by plan of redemption (P) is followed by the cross (though in his foreknowledge God's plan of redemption is sometimes said to be prior to the creation event as well). The cross is thus turned into the final proof in a theorem, the first variable of which does not include or require the God of the cross at all except perhaps through an additive process (C + F + P = †). For orthodox Christians this is surely a grave theological problem.

Literalists will respond that their approach is the only one that preserves the classical doctrine of the atonement. Hence the title to one creationist book, which boldly wagers the entire significance of Christ's life, death and resurrection not simply on the duration of the days of Genesis but on the fathoms deep of Noah's deluge: *Creation, Catastrophe, and Calvary: Why a Global Flood Is Vital to the Doctrine of Atonement.* But while these ways of relating the New Testament to the Hebrew Bible might have a certain simplifying clarity for many believers, they also reflect a highly questionable set of assumptions about the narrative arc of Scripture. They fail to see (or refuse to acknowledge) that strictly penal-substitutionary readings of Christ's death and resurrection rest upon a relatively late and individualistic turn in Christian thinking, replacing a more ancient tradition of "ransom" or *Christus Victor* theology that emphasized not human "genetic" sinfulness but rather Christ's cosuffering and copresence with all of creation, and his battling against and gaining victory over powers holding all finite creatures in bondage to decay. Such a ransom theology is clearly amenable to evolutionary frameworks in ways the individualistic legal-forensic model is not.[3]

God's way of creating, in this understanding, cannot be separated from God's way of redeeming and never could be separated from the beginning. God creates as he redeems and redeems as he creates so that the two are always part of the same act, C† or †C. But what if we will never understand either Genesis or natural history properly if we do not begin with a radically Christocentric understanding of the *character* of God and the *governance* of

God as revealed in the Jesus of history who is the crucified Savior of the world? This is the possibility that kenotic theology would have us wrestle with—that what literalists have long charged is theistic evolution's greatest weakness is in fact its greatest strength. As Polkinghorne writes:

> Christian theology has never simply equated God with Jesus, nor supposed that the historic episode of the incarnation implied that there was, during its period, an attenuation of the divine governance of the universe. The incarnation does, however, suggest what character that governance might at all times be expected to take. It seems God is willing to share with creatures, to be vulnerable to creatures, to an extent not anticipated by classical theology's picture of the God who, through primary causality, is always in total control. . . . [I]n allowing the other to be, God allows creatures their part in bringing about the future.[4]

This response to the problem of animal suffering and "natural evil" will of course be hard for believers in conservative wings of the Reformed tradition to accept. Christians who insist that God's omnipotence entails his absolute predestination of all events, including even human choices, will see little reason to grant nature any space of authentic freedom or indeterminancy either. Some Barthians who insist upon an unbridgeable chasm between God and his creation will also struggle with Polkinghorne's embrace and reformulation of the task of "natural theology." I have no stake in defending such pictures of God. Whatever its difficulties, the only position that makes any moral, religious or rational sense of human moral evil to my mind is the one that declares that the divine will *wills* human free will, and is both powerful enough and self-giving enough to create beings with the capacity to make meaningful, self-defining choices that are morally and spiritually significant. And in the same way we speak of moral evil as resulting from human free will, we should now somewhat analogously speak of natural evil and animal suffering as emerging from free or indeterminate processes, which God does not override and which are inherent possibilities in a creation in which the Creator allows the other to be truly other. God grants the creation the freedom of its own being. "The Creator wills that his creation itself should affirm and continue his work," writes Dietrich Bonhoeffer, "he wills that created things should live and create further life."[5] And God continues to create *in* and *through* these processes while still allowing the creation to be as it is, each element and organism working out its inner principles according to its kind.

The Creator God revealed in the *kenosis* of Christ is neither the remote Designer or Grand Engineer (*deus otiosus*) of Enlightenment deism, nor what Polkinghorne calls the "Cosmic Tyrant" of classical theism who utterly dominates animals not simply once but twice, first in the act of forming them without allowing them to participate in their own making, and second in the act of cursing them without granting them any understanding of their own suffering. Instead, a kenotic picture of the Creator insists that God's creative might and sovereign rule are always expressed in harmony with his character as revealed in the historical person of Jesus, whose way was one of cosuffering humility, nonviolent self-limitation and liberal self-donation. As John Haught writes, a christocentric theology that places such a high premium on creaturely freedom awakens us not so much to the *design* of creation as to its *drama*. The world that God calls into being does not have the character of a "perfect" contrivance or complex invention to be disassembled using techniques of reverse engineering so as to prove God's existence (in the manner of "intelligent design" theory). A god who could be so trapped beneath a microscope would not be the self-revealing and self-concealing God of Jewish and Christian faith at all. Rather, the creation is best seen as an improvisational theater or musical performance in which the director invites the actors—and not human actors alone—to join in the writing of the script, with all of the danger and all of the possibility that this implies. "A God of freedom and promise invites, and does not compel, the creation to experiment with many possible ways of being, allowing it to make 'mistakes' in the process," Haught writes. "This is the God of evolution—one who honors and respects the indeterminancy and narrative openness of creation, and in this way ennobles it."[6]

Or as Terence Fretheim writes of "natural evils" such as earthquakes and floods, "the created moral order" is best grasped as "a complex, loose causal weave." God "lets the creatures have the freedom to be what God created them to be." At the same time, "the looseness of the causal weave allows God to be at work in the system in some ways without violating or (temporarily) suspending it."[7] This opens the door to the possibility of suffering, whether from the sheer randomness of plate tectonics and bolts of lightning that set forests ablaze or from the rise of adaptations in some creatures that are harmful to others. We might summarize this view of the natural world (although, as Cunningham points out, theologically all natural/supernatural dualisms are problematic and only defensible from the standpoint that the creation is supernatural and God alone *natural*[8]) by saying that God's way of creating and sustaining primarily takes the form of divine *providence* working within history, including natural history, rather than absolute *miracle* radically interrupting history from without (which is by no means to deny the possibility of what to human eyes might appear as "interrupting" miracles in other contexts, or even as punctuating parts of the creation process/event itself).

Such a paradigm of creation fits well, some have found, not only with the evidences of biology and geology—helping to make both theological and scientific sense of those unsettling parts of nature creationists seldom care to linger upon—but also with the cosmology of the new quantum physics. In place of the old billiard ball model of causation in Newtonian physics, and even contra Einstein, who attributed all seeming indeterminancy to our incomplete knowledge of the processes at work ("God does not play dice," Einstein famously declared), the quantum factor of the new physics says that there is *real* indeterminancy in the universe, that at the most basic level of existence—the level of elementary particles and atomic structures—there is radical uncertainty so that there can even be effects without causes. The theological implications of Heisenberg's celebrated uncertainty principle are as disturbing to the Designer God of classical theology as Darwin's theory of natural selection. Is there not something defective or weak or negligent, we might well ask, in a Creator who would inscribe such lawlessness, such lack of predetermined order, at the very heart of material existence? Or is it in fact we ourselves who have long held defective notions about God's character, which must be completely rethought in the light of the self-emptying Christ of the New Testament—the One who draws all of creation ever deeper into his own fullness of life with an implacable yet noncoercive and infinitely patient love, the King who scandalously creates and rules the universe from a throne in the form of a cross? And are we prepared to follow this Creator who neither prevents nor rationalistically explains but instead *enters into* the suffering and contingency of his creation and in so doing redeems it?

There is still another sense in which we must learn to read Genesis in radically christocentric theological terms rather than as mere historical chronicle. For orthodox Christianity, Cunningham points out, it is not Adam but *Christ* who is the first true human, the *axis mundi* by whom we must now reenvision all that came before as well as all that comes after. Some have insisted that without a historical Adam the life, death and resurrection of the historical Jesus would be devoid of meaning. But this claim amounts to a denial (even if unintentionally so) of the centrality of Christ; for it gives the fallen Adam of Genesis an interpretive primacy over the Jesus of history that Paul and the Gospel writers do not allow. For disciples of Christ, it is only *in Christ* that the ancient story of human origins and destiny can be rightly understood—not the other way around. We do not read the story of Christ "Adamically." We reread the story of Adam *christologically* in the light of the second Adam who is also the *first* Adam, the first fully human being of whom the ancient story is only a type, a dim shadow and longing, a "figure of him that was to come" (Rom 5:14). In the Common English Translation, those passages in the Gospels in which Christ refers to himself as "the Son of Man" are translated "the Human One." The New Testament proclamation is not that the Adam of Hebrew Scripture must now be greatly elevated as the father of humankind lest Christ have died a pointless death. It is that He who comes last is first. The Christian *euangelion* is not an accentuation or amplification but, in a real sense, a subversion of the first Adam's theological and historical significance (whether or not a historical Adam existed). It is only through the *kenosis* of Christ—his self-emptying death upon a "tree"—that our eyes have at last been opened to the real nature of good and evil for the first time. The cross is at once the two trees in the Garden of Eden, the tree of knowledge and the tree of life. When Christ cries, "It is finished" on Easter Friday, the creation of the world is at last completed. Nor is Christ's rest in the tomb an observance of Jewish sabbath law. It is the *first* sabbath to which Jewish law and the creation story proleptically pointed. Genesis is not science or journalism but *prophecy*. And it is by entering into Christ's way of self-emptying love and reposing with him in his sabbath rest that we bear witness to this hope: that one day we will also share in our Lord's resurrection and glorification. Only then will Christ be all in all. The sabbath, as Cunningham writes, "is therefore the meaning of creation"—we are "a species of the sabbath."[9]

14

Animal Ethics, Sabbath Rest

❧

When I was a child my missionary family moved three times, from Thailand to Taiwan to Zimbabwe. When people would ask me where I was from, I never knew exactly how to answer them (and still don't to this day). My passport said I was from the United States, but I had little knowledge of life in America, which we only visited for a few months every four or five years and which was more exotic to me than Bangkok, Taipei or Harare. Yet I wasn't "from" any of these places either. I learned from an early age to be at ease everywhere but at home nowhere. I was not completely rootless, however, for my family kept the sabbath, and this, I have come to realize, was my real childhood home—not a place but a time as real as any physical locality. Every Friday my sisters and I would spend the day (not always cheerfully, one must confess) helping my parents clean the house so that it would be "ready for the sabbath." My father, a single-task man, would carefully spread out old newspapers on the back porch on which to shine his shoes. My mother would meanwhile be stripping sheets off beds, scrubbing bathtubs, vacuuming carpets and baking all at once. Time accelerated on Fridays. There was so much to be done to prepare for the sabbath, which would come the moment the sun dipped below the horizon, like an honored guest in a magical fable, whether we had prepared to welcome it or not. It would be a terrible shame for the sun to set and for the sabbath to arrive unnoticed in homes and with lives still in careless disarray.

When the sabbath came, however, time slowed down. Something in the fabric of reality changed. My father selected "sabbath music"—choral works or classical masters on the cassette tapes and records we carried with us in boxes from continent to continent. We lit sabbath candles. Every Friday my mother baked fresh cinnamon rolls to welcome the sabbath. I do not know whether Christians who have never greeted the sabbath in the warm glow of candles or with the aroma of cinnamon rolls have truly encountered the day. I do know that there are mysteries of life that can only be understood through patterns of sacramental living that reenchant the world and mend the frayed strands of existence. To be a sabbath-keeper is to be a participant in such a sacramental drama.

Yet the thrilling romance of sabbath keeping (to borrow from the words of G. K. Chesterton) is not only for mystics at heart. In Hebrew Scripture, the meaning of the day has always been eschatological in nature. It has always been oriented most strongly toward the future rather than the past or even the present. Anxious notions about the need for one-to-one correspondence between covenantal living and the timeframes of Genesis 1 defy all wisdom. They are alien to Jewish thinking, which has embraced nonliteralistic interpretations of Genesis for centuries if not millennia—from the most revered interpreter of the Torah in Jewish history, Moses Maimonides in the twelfth century, to the ultraorthodox Rabbi Kook in the twentieth.[1] The creation "is not an act that happened once upon a time, once and forever," writes Hasidic rabbi and process theologian Abraham Joshua Heschel in his classic meditation on the meaning of the sabbath in Jewish thought. "The act of bringing the world into existence is a continuous process. . . . Time is perpetual innovation, a synonym for perpetual creation."[2] To keep the sabbath, then, is to remember that the Lord God is the One who creates not simply once but every unfolding moment of our lives. And it is to *remember what lies ahead*. The sabbath, in Heschel's indelible words, is a "palace in time"—not a stone monument to a frozen past but an intimation of eternity. The very word *keep* in the sabbath commandment in the book of Deuteronomy may also be translated "eagerly await."[3] We eagerly await the sabbath. But the sabbath day is itself a day of eager awaiting that anticipates something greater. To be a sabbath keeper is to cultivate an awareness of eternity through the spiritual discipline of situating one's life within a forward-moving narrative of God's unfolding purposes for all of his creation. These purposes are purposes of compassion, mercy and justice that become woven into the fabric and pulse of our daily lives as we enter into the weekly *shalom* of the divine rest. To *Christianly* keep the sabbath is to eagerly await the Second Coming, to live amid the shadows in a way that embodies the hope that this is only the interval—the liminal space—between cross and resurrection.

The sabbath first appears in the Hebrew Bible not as a commandment (and not by name) but as an implicit invitation in the Genesis story for humans made in the image of God to realize their full humanity by joining with the Creator in abstaining from toil. The fact that the sabbath marks the first full day of the human pair's existence in the Genesis narratives suggests that in the divine economy rest is not for the sake of labor but labor for the sake of rest. God does not give humans rest only after they have "earned" it through the sweat of their brows. He gives it to them "on arrival." Hence the sabbath is sheer gift. It is also the sign and sacrament of true human communion with one another and with their Creator.

The first time the sabbath is mentioned by name is in the book of Exodus. After fleeing their Egyptian slave masters, the Israelites find themselves without food in the wilderness. God miraculously provides manna for them. The Israelites are instructed by God to gather enough of the bread from heaven for their daily needs but not more (Ex 16:16-18), to not hoard or store up the divine gift (Ex 16:19-20), and to gather twice as much on the sixth day, for "on the seventh, which is the Sabbath, in it there shall be none" (Ex 16:26). Those who attempt to collect more than their daily bread find that what they have greedily or fearfully accumulated spoils, while those who take only according to their needs find that their manna is preserved. The manna parable in Exodus, which includes the first clear prescription in the Bible of how to *keep* the sabbath holy, is therefore a story of sabbath economics. In contrast to Egypt's "military-industrial" economy built upon wealth accumulation and exploitation of labor, the divine economy rests upon principles of abundance, sufficiency, limited human needs, and dependence upon the Creator God who provides for all.[4] The divine *Enough!* in the face of human attempts to own God's gifts and turn them to private advantage is a lesson aimed at liberating the Israelites from the practices of anxious, acquisitive exploitation of the earth and of other human beings.

There is not only a sabbath day in Jewish law, however, but also a sabbath year and a sabbath year of years, the Jubilee, which the weekly sabbath foreshadows and in a limited way participates in. Every seventh year, according to the book of Deuteronomy (Deut 15), all debts must be forgiven. Sabbath economics includes a recurring structural adjustment policy aimed not at extracting the last penny owed through crushing austerity measures, but rather at breaking the cycle of disequalizing wealth concentration, thus restoring social justice and political balance. But even the sabbath year turns out to be only a very partial glimmering of what lies ahead.

The year of Jubilee or "sabbath of sabbaths," commanded to occur every forty-nine or fifty years, includes not only the cancellation of all debts but also the return of all land to its original owners and the release of all slaves *with gifts*. Although historians have questioned whether the Jubilee laws were ever fully enacted by the Israelites, they were repeatedly appealed to by the Hebrew prophets as a measure of how faithfully the nation was keeping the weekly sabbath. Amos rebukes Israel's wealthy classes for betraying the sabbath in their relentless quest for profit on the backs of the poor (Amos 8:5-6). Jeremiah issues a scathing critique of King Zedekiah for reneging on his promises to enact the Jubilee (Jer 34:13-16). Nehemiah condemns the practice of loaning with interest and calls Israel back to the radical countervision of sabbath economics (Neh 5:6; 10:31; 13:15-22). Ezekiel reminds Israel's princes of the "year of liberty" and warns against the injustices of eviction and dispossession (Ezek 46:17-18; 47:13-23). The prophet Isaiah delivers the following searing message to those self-satisfied sabbath worshipers whose scrupulous piety without any concern for violence and oppression is nothing more than a noxious affront to the Lord:

> New moon and Sabbath, the calling of assemblies—
> I cannot endure iniquity and the solemn assembly.
> I hate your new moon festivals and your appointed feasts,
> They have become a burden to Me;
> I am weary of bearing them.
> So when you spread out your hands in prayer,
> I will hide My eyes from you;
> Yes, even though you multiply prayers,
> I will not listen.
> Your hands are covered with blood.
> Wash yourselves, make yourselves clean;
> Remove the evil of your deeds from My sight.
> Cease to do evil,
> Learn to do good;
> Seek justice,
> Reprove the ruthless,
> Defend the orphan,
> Plead for the widow. (Is 1:13-17 NASB)

Isaiah's inspired poetics end with a vision of the sabbath Jubilee—"the favorable year of the LORD"—on an eschatological horizon that lies ahead and that will come as "good news to the afflicted" and "freedom to prisoners . . . for I, the Lord, love justice" (Is 61:1-2, 8 NASB).

But there is still more to the ethics of radical generosity signified by the sabbath. In a statement of profound ecological concern, Exodus 23 extends the gift of the sabbath to the nonhuman world and even to the land. Every seventh year, the Israelites are commanded, "You shall let [the land] rest and lie fallow, so that the needy of your people may eat; *and whatever they leave the beast of the field may eat*" (Ex 23:11 NASB). Here the productivity and fruitfulness of the earth is dramatically linked to human noninterference in the natural world as a regular corrective to human subduing. Care for the land is also linked by the Creator to care for the socially marginalized and even for nondomesticated creatures. This command of free trespass may be the earliest law of wild animal rights in human history, closely following on the commandment in the Decalogue that domesticated animals shall not be made to work on the sabbath (Ex 20:10).

While Jewish concern for animal well-being may be read as a holistic extension of the principles of sabbath observance, rabbinic teaching in ancient Israel permitted the breaking of the sabbath to alleviate the suffering of animals. Jews may unload heavy burdens from animals on the sabbath, for example, so that they are not left in discomfort, an activity normally forbidden. The sabbath was made for animals, not animals for the sabbath. In the New Testament, the Rabbi Jesus thus defends his ministry of healing on the sabbath by reminding those who have charged him with playing fast and loose with the holiness of the day that authentic sabbath keeping is, first and foremost, a summons of justice and compassion that extends even toward animals. "What man shall there be among you, that shall have one sheep, and if it fall into a pit on the sabbath day, will he not lay hold on it, and lift it out?" (Mt 12:11-12). Other Jewish laws focused on animal welfare include the command that oxen must not be muzzled as they plough but should be allowed to eat freely (Deut 25:4); and "If you see the donkey of one who hates you lying helpless under its load, you shall refrain from leaving it to him, you shall surely release it with him" (Ex 23:5 NASB). Jewish law forbids the killing of cows or sheep on the same day as their young, or the separation of newborn calves and lambs from their mothers within the first week of their birth (Lev 22:27). It requires that a mother bird be sent away before its eggs are taken to avoid causing it undue distress (Deut 22:6-7). Jewish tradition and Talmudic teaching condemn hunting for sport; the only two hunters mentioned in the Bible, Nimrod and Esau, are both depicted in an unsavory if not villainous light.

The theological and moral meanings, textures and resonances of the sabbath day have been almost entirely forgotten by most Christians. I do not say this as a polemic against the tradition of worshiping on the Lord's Day in commemoration of Easter Sunday. The best historical evidence suggests that from a very early period Christians simultaneously kept the sabbath or Holy Saturday and remembered Christ's resurrection on the first day of the week. The practice of sabbath keeping was not *transferred* to Sunday. Instead, it was gradually forgotten if not consciously abandoned by Gentile Christians seeking to distance the new faith from its Jewish heritage. But the theological meanings of the two days are not the same, and there are good reasons for Christians of all traditions to remember and reclaim the biblical sabbath, including, in a post-Holocaust world, as an act of repentance for centuries of anti-Semitism. Would the Shoah have been conceivable had Christians from the beginning remembered and kept the sabbath in all of its sacramental richness and moral significance as central to the meaning of Christ's life? And if we now face an ecological holocaust in which manic human greed and unchecked exploitation of the earth threaten to destroy entire species and render God's creation a wasteland, might not the holistic moral vision signified by the discipline of sabbath rest restore lost sanity?

For Christians, Jesus may be more than the culmination of the Jewish prophetic tradition, but he is not less. Christ begins his ministry in the Gospel of Luke by entering a synagogue on the sabbath day and reading from the prophet Isaiah's Jubilee announcement of good news to the poor. After closing the scroll, Jesus declares, "Today this Scripture has been fulfilled in your hearing" (Lk 4:21 NASB). Christ thus boldly declares that he has come not to abolish but to inaugurate and restore the deepest meaning of the sabbath, not from any preoccupation with literalistic hermeneutics or "scientific" creationism, but rather as a summons to *ethical* creationism. This accent on the moral dimensions of the sabbath lies at the heart of the Hebraic tradition as well as the earliest Christian memories of the Jesus of history.

Tragically, biblical literalists and creationists—and indeed, Christians in general—have not remembered the radical economics of the sabbath Jubilee with its summons to wealth redistribution and justice to the poor. Nor have they been at the forefront of the environmental or the animal-rights movements. Most Christian organizations and individuals, Wendell

Berry writes, "are as happily indifferent to the ecological, cultural and religious implications of industrial economics as are most industrial organizations. The certified Christian seems just as likely as anyone else to join the military-industrial conspiracy to murder creation."[5] There may in fact be a fateful connection between longstanding Christian readings of Genesis and Christian complicity in the destruction of the planet, including the infliction of terrible new forms of unnatural cruelty visited upon animals on a staggering scale in the modern age. "It is almost a chorus among commentators that God created the world alone, with overwhelming power and absolute control, while working independently and unilaterally," observes Fretheim:

> But if this understanding of the imaging of God in Genesis 1 is correct, then those created in God's image could properly understand their role regarding the rest of creation in comparable terms: in terms of power over, absolute control, and independence. By definition, the natural world thus becomes available for human manipulation and exploitation.[6]

Why, after all, should believers care about the nine billion animals butchered annually in the United States—the cattle routinely dismembered alive, the hogs plunged still conscious into vats of boiling water, the birds packed so tightly into cages to be trucked thousands of miles that they often arrive crushed and suffocated on delivery—if the God we worship is a God whose creation is simply a mode of "stamping" animals into existence by verbal decree before delivering them over to humans for their instrumental "dominion"/consumption? Why should Christians care about the abuse inflicted every second of every day upon sentient creatures in slaughterhouses around the world if the divine benediction was immediately superseded by a divine malediction or "curse" upon all animals? And if the world is bound for a fiery conflagration in the near future in which all animals will be destroyed by God anyway, as many fundamentalist Christians believe, why should we care about their suffering in the present or invest our time and energy in alleviating their pain? Shouldn't we instead devote ourselves to "evangelism" (ignoring the fact that the greatest likely cause of planetary destruction is not divine intervention but the rapaciousness of human beings themselves, or that the *euangelion* of Christ in the New Testament is a summons to *dikaiosyne,* which means not simply righteousness but *justice*)? Or again, if humans have no intimate familial relationship with the rest of the animal world, why shouldn't human "subduing" take the form of unrestrained predation and violence upon other creatures?

Even if there is no inevitable link between biblical literalism and indifference to animal suffering, creationists must face the fact that while they have been engrossed with proving the scientific correctness of their theology (or with constructing creationist-themed amusement parks or with fighting costly legal battles in vain attempts to insert their religious beliefs into public high school classrooms), evolutionary biologists and naturalists—including avowed atheists—have taken the lead in actually protecting God's creation. If there is to be a final day of reckoning in which the living and the dead shall be called to give an account of what they did with their lives, *sub specie aeternitatis,* Jane Goodall, Louis Leakey and Rachel Carson, not George McCready Price, may one day find themselves being commended by the Almighty for having been faithful *creationists*—for having fought to protect the lives of animals when others who loudly claimed to be God's chief spokespersons viewed the task of actually *caring* for the creation with reluctance, nonchalance or outright disdain.

"There is nothing more animal-like than a clear conscience on the third planet of the Sun," writes Polish Nobel laureate Wislawa Szymborska.[7] It is a morally ambiguous verse, and some readers might imagine that an animal-like clear conscience would be a fine thing for humans to possess. According to Nietzsche, the highest human types have always acted without concern for the weak, with light hearts like the panther and the killer whale in Szymborska's poem. They "appear as lightening appears," Nietzsche declares. "Their work is an instinctive creation and imposition of forms; they are the most involuntary, unconscious artists there are."[8] And the unconscious artistry of the Übermenschen includes violence without guilt. But while there may be moral innocence in animal predation and even a terrible beauty in nature's mysterious passion play of life-giving death—a beauty that is perhaps the closest we can hope to come to answering the theodicy riddle of animal suffering—there is no innocence in human violence toward other created beings. When those uniquely made in the divine image inflict unnecessary pain and death upon other sentient creatures for the sake of their own pleasure or profit, they are rightly called not superhumans but *sub*-humans and the *inhumane*.

Herein, it seems to me, lies the most pressing theological dilemma of our age—not the theodicy dilemma of evolutionary biology but the *anthropodicy* dilemma of late capitalism. Is it still possible to justify the existence of that species that has become a force of such destruction on the planet that it is no longer clear that other species will survive? Does the *imago Dei* remain, or shall we devour the earth that was left in our care without restraint until it is an utterly scorched desert? Any credible answer to these questions, which grow in urgency every day, must take the form not of detached theologizing but of concrete and ethical action that brings sabbath peace to our brothers and sisters in the animal kingdom. Their blood may or may not be upon God's hands. On the third planet of the sun there can be no doubt that it is now upon our own.

Conclusion

Having warned throughout this book of the destructive power of rigidly literalistic or fundamentalist approaches to Genesis, both at the level of individuals and of faith communities, I do not want to end without sounding a cautionary note concerning "fundamentalisms" of a different kind. One of the reasons I became convinced that there is something deeply unhealthy gnawing at the heart of biblical literalism on Genesis and "scientific" creationism is because of the toxic speech habits, blatant power maneuvers (wrapped in exalted religious rhetoric) and tactics of misrepresentation and incrimination I have witnessed in some of the movement's most vocal defenders within my own tradition. Those who manifest such uncharitable spirits, I concluded early on in my wrestling with questions of faith and science, are probably not the divinely anointed guardians of truth.

At the same time, I have often been equally dismayed by the attitudes evinced by some individuals on the other side of the debate over creation and evolution—by how quickly some are prepared to write off people of sincere faith who are at different places in their intellectual and spiritual journeys, by how little pastoral sensitivity they show in introducing unsettling ideas to others, and by how often their sense of scientific certainty and mastery of technical knowledge assumes the character of its own ersatz religion. Some of these individuals, I have concluded, have indeed broken faith with their Christian communities, though not for reasons as facile as whether or not they affirm literalistic dogmas. Because there are no easy answers to the dilemma of animal suffering, greater charity and greater civility is required of all of us. I have tried my best in this book to vigorously challenge what I take to be damaging and myopic ideas about Scripture without maligning the characters of individual people who hold them. Still, I recognize I have much to learn and unlearn in dialogue with other Christians, including literalists, creationists and fundamentalists if they desire to be in an authentic dialogue with believers who have come to different conclusions than they have about the meaning of Genesis.

I will end with a parable. In Flannery O'Connor's 1965 short story "Revelation," the obese, racist, fundamentalist southern farmer's wife Mrs. Turpin has just returned to her hog farm after visiting the doctor's office, where a sullen, acne-faced teenaged girl in the waiting room named Mary Grace mysteriously and violently hurled a book—titled *Human Development*—into her face. Mrs. Turpin is suddenly having a crisis of faith. As she sits outside her pigpen indignantly questioning how God could have allowed her—her of all people!—to be humiliated in such an undignified way, she receives a vision:

> There was only a purple streak in the sky, cutting through a field of crimson and leading, like an extension of the highway, into the descending dusk. She raised her hands from the side of the pen in a gesture hieratic and profound. A visionary light settled in her eyes. She saw the streak as a vast swinging bridge extending upward from the earth through a field of living fire. Upon it a vast horde of souls were rumbling toward heaven. There were whole companies of white-trash, clean for the first time in their lives, and bands of black niggers in white robes, and battalions of freaks and lunatics shouting and clapping and leaping like frogs. And bringing up the end of the procession was a tribe of people whom she recognized at once as those who, like herself and Claud, had always had a little of everything and

> the God-given wit to use it right. She leaned forward to observe them closer. They were marching behind the others with great dignity, accountable as they had always been for good order and common sense and respectable behavior. They alone were on key. Yet she could see by their shocked and altered faces that even their virtues were being burned away. She lowered her hands and gripped the rail of the hog pen, her eyes small but fixed unblinkingly on what lay ahead. In a moment the vision faded but she remained where she was, immobile.
>
> At length she got down and turned off the faucet and made her way on the darkening path to the house. In the woods around her the invisible cricket choruses had struck up, but what she heard were the voices of the souls climbing upward into the starry field and shouting hallelujah.[1]

O'Connor spares nothing in her diagnosis of Mrs. Turpin's spiritual and mental sickness unto death. But God has a revelation even for Mrs. Turpin. Despite all of her faults and obvious human limitations, she is privileged to receive a divine vision of penetrating clarity and ecumenical moral significance. No human being—and no community either—is outside the fold of God's concern or God's violent, disrupting grace. So we must heed the cunning trap O'Connor has set for us. The entire time the fundamentalist, uneducated, foolish Mrs. Turpin has been sitting in the waiting room judging the rest of its occupants with by turns comical and grotesque spiritual pride, we "progressive" and enlightened readers who hold all of the correct views about human development have been sitting in judgment on her. I take this to mean that in the divine comedy of God's final judgment we must also one day join Mrs. Turpin at the back of the line to have even our virtues burned away.

Notes

Introduction

[1]Georg W. F. Hegel, *The Philosophy of History*, trans. J. Sibree (New York: Prometheus Books, 1991), p. 21.

[2]David Bentley Hart, *The Doors of the Sea* (Grand Rapids: Eerdmans, 2005), p. 50.

[3]Anne Dillard, *The Annie Dillard Reader* (New York: HarperCollins, 1994), p. 374.

[4]Ibid., p. 318.

[5]Terence E. Fretheim, *Creation Untamed: The Bible, God, and Natural Disasters* (Grand Rapids: Baker Academic, 2010), p. 3.

[6]Clifford Goldstein, "Seventh-day Darwinians," *Adventist Review* (July 24, 2003): 29.

[7]On the influence of Seventh-day Adventism on American creationism see, for example, Mark Noll, *The Scandal of the Evangelical Mind* (Grand Rapids: Eerdmans, 1995), pp. 189-94; and Ronald Numbers, *The Creationists: From Scientific Creationism to Intelligent Design* (Cambridge: Harvard University Press, 2006).

[8]Frank Newport, "Four in 10 Americans Believe in Strict Creationism," Gallup Poll (December 17, 2010), www.gallup.com/poll/145286/four-americans-believe-strict-creationism.aspx.

[9]John H. Walton, *The Lost World of Genesis One: Ancient Cosmology and the Origins Debate* (Downers Grove, IL: IVP Academic, 2009), p. 110.

[10]John Haught, "The Darwinian Struggle," *Commonweal* 106, no. 16 (September 24, 1999): 14-16.

[11]See "The Trial of God" in my book *Anarchy and Apocalypse: Essays on Faith, Violence and Theodicy* (Eugene, OR: Wipf and Stock/Cascade Books, 2010).

Chapter One

[1]Dietrich Bonhoeffer, *Creation and Fall: Two Biblical Studies* (New York: Touchstone, 1959), p. 50.

[2]Karl Barth, *Church Dogmatics: A Selection with Introduction by Helmut Gollwitzer* (Edinburgh: T & T Clark, 1961), p. 350.

[3]Robert Alter, *Genesis: A Translation and Commentary* (New York: W. W. Norton, 1996), p. 9.

[4]John Locke, *Second Treatise of Government* (Indianapolis: Hackett Publishing, 1980), p. 21.

[5]Mark S. Whorton, *Peril in Paradise: Theology, Science, and the Age of the Earth* (Downers Grove, IL: InterVarsity Press, 2005), pp. 195-96.

[6]Andrew Linzey, *Animal Theology* (Chicago: University of Illinois Press, 1994), p. 81.

[7]Walter Brueggemann, *Genesis* (Atlanta: Westminster John Knox, 1982), p. 42.

[8]David Charles Kraemer, *Responses to Suffering in Classical Rabbinic Literature* (Oxford: Oxford University Press, 1995), pp. 135-36.

[9]See, for example, Franz Delitzsch, *A New Commentary on Genesis* (Edinburgh: T & T Clark, 1888), p. 160.

[10]Alter, *Genesis*, p. 13.

[11]Alan E. Lewis, *Between Cross and Resurrection: A Theology of Holy Saturday* (Grand Rapids: Eerdmans, 2001), p. 409.

[12]John Stott, *Understanding the Bible: Expanded Edition* (Grand Rapids: Zondervan, 1999), pp. 54-55.

[13]John H. Walton, *The Lost World of Genesis One: Ancient Cosmology and the Origins Debate* (Downers Grove, IL: IVP Academic, 2009), p. 21.

[14]See Alister McGrath, *A Fine-Tuned Universe: The Quest for God in Science and Theology: The Gifford Lectures* (Louisville, KY: Westminster John Knox, 2009).

CHAPTER TWO

[1]See William Brown, *The Seven Pillars of Creation: The Bible, Science, and the Ecology of Wonder* (Oxford: Oxford University Press, 2010).

[2]Gerhard F. Hasel, "The Meaning of the Chronogenealogies of Genesis 5 and 11," *Origins* 7, no. 1 (1980): 57-70.

[3]Merold Westphal, *Whose Community? Which Interpretation? Philosophical Hermeneutics for the Church* (Grand Rapids: Baker Academic, 2009), p. 18.

[4]See Nancey Murphy, *Beyond Liberalism and Fundamentalism: How Modern and Postmodern Philosophy Set the Theological Agenda* (Harrisburg, PA: Continuum/ Trinity Press International, 1996).

[5]René Descartes, "Discourse on Method," in *Descartes: Selected Philosophical Writings* (Cambridge: Cambridge University Press, 1988), p. 36.

[6]Stephen Toulmin, *Cosmopolis: The Hidden Agenda of Modernity* (Chicago: Chicago University Press, 1990), p. 19.

[7]Carl Raschke, *The Next Reformation: Why Evangelicals Must Embrace Postmodernity* (Grand Rapids: Baker Academic, 2004), p. 24.

[8]Henry Morris, "The Literal Week of Creation," Institute for Creation Research, www.icr.org/article/literal-week-creation/.

[9]On creationism as fideism, see Daniel Helminiak, *Religion and the Human Sciences* (New York: SUNY Press, 1998), p. 105.

[10]Hasel, "Chronogenealogies," pp. 69-70.

Chapter Three

[1]As cited in John Whitcomb, *The Early Earth: An Introduction to Biblical Creationism* (Grand Rapids: Baker, 1986), p. 36.

[2]Paul Taylor, *The Six Days of Genesis: A Scientific Application of Genesis 1-11* (Green Forest, AR: New Leaf Publisher, 2007), pp. 128, 135.

[3]Jonathan Sarfati, "Biblical Chronogenealogies," *Journal of Creation* 17, no. 3 (2003): 18.

[4]See John Morris, *The Young Earth: The Real History of the Earth—Past, Present, and Future* (Green Forest, AR: New Leaf Publishing, 1994), p. 30; and Russell Grigg, "Should Genesis Be Taken Literally?" *Creation* 16, no. 1 (1993): 38-41.

[5]Alvin Plantinga, "Evolution, Neutrality, and Antecedent Probability: A Reply to McMullin and Van Till," in *Intelligent Design Creationism and Its Critics: Philosophical, Theological, and Scientific Perspectives* (Cambridge, MA: MIT Press, 2001), pp. 216-18.

[6]Walter Brueggemann, *Genesis* (Atlanta: Westminster John Knox, 1982), p. 25.

[7]Ariel Roth, *Origins: Linking Science and Scripture* (Hagerstown, MD: Review and Herald, 1998), p. 318.

[8]Clifford Goldstein, "Seventh-day Darwinians," *Adventist Review* (July 24, 2003): 29.

[9]Mike Campbell, "Church President Says He Won't 'Flinch' on Creation Issue," *Adventist Review* (July 8, 2010): 11; and "Response to an Affirmation of Creation," voted by the members of the Seventh-day Adventist General Conference Executive Committee at the 2004 Annual Council, October 13, 2004, www.adventist.org/beliefs/statements/main-stat55.html.

[10]Harold W. Clark, *Back to Creationism*, in *The Early Writings of Harold W. Clark and Frank Lewis Marsh*, ed. Ronald Numbers (London: Routledge, 1995), pp. 30, 34.

[11]Jacques Ellul, *Ethics of Freedom* (Grand Rapids: Eerdmans, 1976), pp. 131, 163.

Chapter Four

[1]See, for example, Karl Popper, *Conjectures and Refutations: The Growth of Scientific Knowledge* (London: Routledge, 1963).

[2]Thomas Kuhn, *The Structure of Scientific Revolutions* (Chicago: University of Chicago Press, 1962), p. 147.

[3]Ibid., p. 94.

[4]See Imre Lakatos, "Why Copernicus's Programme Superseded Ptolemy's," in *The Methodology of Scientific Research Programmes: Philosophical Papers, Volume 1*, ed. John Worrall and Gregory Currie (Cambridge: Cambridge University Press, 1978), pp. 172-73.

[5]Arkady Plotnitsky, "Thomas Kuhn," in *Postmodernism: The Major Thinkers* (Oxford: Wiley-Blackwell, 2002), pp. 201-11.

[6]Lakatos, "Falsification and the Methodology of Scientific Research Programmes," in *The Methodology of Scientific Research Programmes*, p. 48.

[7]Some of these creationist-inspired studies published in peer-reviewed scientific journals are described by Leonard Brand in his paper "Insight from Hindsight: Resolving Seeming Conflicts Between Science and Scripture," presented at the Geo-Science Research Institute Conference in Alberta, July 2011, available at http://spectrummagazine.org/print/3344.

[8]See John Polkinghorne, *Faith, Science, and Understanding* (New Haven, CT: Yale University Press, 2000), p. 118.

[9]Ben Clausen, "Nuclear Physics," in *In Six Days: Why Fifty Scientists Choose to Believe in Creation*, ed. John Ashton (Green Forest, AR: Master Books, 2000), p. 273.

[10]Brand predicts, for example: (1) that "the fossil record will eventually reveal that it did not result from evolution of major life forms, as it now appears to"; (2) that "we are now seriously misinterpreting the radiometric data, and it actually gives only relative age, not age in years. . . . We have not resolved the conflict, but we can make a prediction as to what we believe the outcome will be"; (3) that "new archeological evidence will sooner or later support the truth of the biblical account of the Exodus"; and (4) that "ice cores will be found to not be annual layers" (Brand, "Insight from Hindsight").

[11]Survey data shows that the overwhelming majority of scientists who are theists in the United States do precisely this. A 1997 survey published by the National Center for Science Education of more than a thousand biologists, mathematicians, physicists and astronomers found that fully 40 percent were theists who believed in a God who created humans through a purposeful process lasting millions of years, in contrast to five percent believing that God created humans "pretty much in their present form at one time within the last 10,000 years." Larry Witham, "Many Scientists See God's Hand in Evolution," *Reports of the National Center for Science Education* 17, no. 6 (November–December 1997): 33.

CHAPTER FIVE

[1]As cited in Ronald Numbers, *The Creationists: From Scientific Creationism to Intel-*

ligent Design (Cambridge, MA: Harvard University Press, 2006), p. 105.

[2]Leonard Brand, "Insight from Hindsight: Resolving Seeming Conflicts Between Science and Scripture," paper presented at the Geo-Science Research Institute Conference in Alberta, July 2011, http://spectrummagazine.org/print/3344.

[3]See Nancey Murphy, "Phillip Johnson on Trial: A Critique of His Critique of Darwin," in *Intelligent Design Creationism and Its Critics: Philosophical, Theological, and Scientific Perspectives,* ed. Robert T. Pennock (Cambridge, MA: MIT Press, 2001), p. 464.

[4]G. K. Chesterton, *The Everlasting Man* (San Francisco: Ignatius, 1993), p. 25.

Chapter Six

[1]As cited in Ronald Numbers, *The Creationists: From Scientific Creationism to Intelligent Design* (Cambridge, MA: Harvard University Press, 2006), p. 92.

[2]Oliver O'Donovan, *Church in Crisis: The Gay Controversy and the Anglican Communion* (Eugene, OR: Cascade Books, 2008), p. 4.

[3]Karl Barth, *Karl Barth–Rudolph Bultmann Letters, 1922–1966* (Grand Rapids: Eerdmans, 1981), p. 145.

[4]Emmanuel Sivan, "The Enclave Culture," in Gabriel A. Almond, R. Scott Appleby and Emmanuel Sivan, *Strong Religion: The Rise of Fundamentalisms Around the World* (Chicago: University of Chicago Press, 2003), p. 56.

[5]Edward Farley, "Fundamentalism: A Theory," *Crosscurrents* 55, no. 3 (Fall 2005): 397.

[6]See Erik Erikson, *Identity and the Life Cycle, Volume 1* (New York: W. W. Norton, 1980); and John Van Wicklin, Ronald J. Burwell and Richard E. Butman, "Squandered Years: Identity Foreclosed Students and the Liberal Education They Avoid," in *Assessment in Christian Higher Education: Rhetoric and Reality,* ed. D. John Lee and Gloria Goris Stronks, (Lanham, MD: University Press of America, 1994), pp. 79-89.

[7]John Milbank, *Theology and Social Theory: Beyond Secular Reason* (London: Blackwell, 1990), p. 21.

[8]N. T. Wright, "How Can the Bible Be Authoritative?", The Laing Lecture and the Griffith Thomas Lecture, 1989, www.ntwrightpage.com/Wright_Bible_Authoritative.htm.

[9]Ibid.

Chapter Seven

[1]C. S. Lewis, "The Inner Ring," in *The Weight of Glory* (San Francisco: HarperCollins, 1949), p. 147.

[2]For a broad survey of gnostic origins, beliefs and practices, see Hans Jonas, *The Gnostic Religion* (Boston: Beacon Press, 1958).

[3]Ibid., p. 104.

[4]Ibid., pp. 320-40.

[5]Eric Voegelin, *The Collected Works of Eric Voegelin, Volume Five: Modernity Without Restraint* (Columbia, MO: University of Missouri Press, 2000).

[6]Luciano Pellicani, *Revolutionary Apocalypse: The Ideological Roots of Terrorism* (London: Praeger, 2003), p. 151.

[7]Ibid.

[8]Ibid., pp. 151-52.

[9]Norman Cohn, *The Pursuit of the Millennium: Revolutionary Millenarians and Mystical Anarchists of the Middle Ages* (London: Pimlico, 2004), p. 284.

[10]Ibid., pp. 202-3.

[11]Voegelin, *The New Science of Politics*, in *Modernity Without Restraint*, p. 226.

[12]Pellicani, *Revolutionary Apocalypse*, p. 204.

[13]Ibid., p. 212.

[14]Ibid., p. 213.

[15]Jonas, *Gnostic Religion*, p. 146.

[16]Conor Cunningham, *Darwin's Pious Idea: Why the Ultra-Darwinists and the Creationists Both Get It Wrong* (Grand Rapids: Eerdmans, 2010), p. 23.

CHAPTER EIGHT

[1]Conor Cunningham, *Darwin's Pious Idea: Why the Ultra-Darwinists and the Creationists Both Get It Wrong* (Grand Rapids: Eerdmans, 2010), p. 378.

[2]As cited in David Deane, *Nietzsche and Theology: Nietzschean Thought in Christological Anthropology* (Hampshire, UK: Ashgate Publishing, 2006), p. 158.

[3]George Hunsinger, *How to Read Karl Barth: The Shape of His Theology* (Oxford: Oxford University Press, 1991), p. 44.

[4]Karl Barth, *Church Dogmatics: Volume 3, Part 1: The Doctrine of Creation* (London: T & T Clark, 1958), pp. 81-86.

[5]John Calvin, *Genesis*, ed. Alister McGrath and J. I. Packer (Wheaton, IL: Crossway Books, 2001), p. 19.

[6]Ibid., p. 30.

[7]Peter M. van Bemmelen, "Divine Accommodation and Biblical Creation: Calvin vs. McGrath," *Andrews University Seminary Studies* 39, no. 1 (2001): 109-16.

[8]Calvin, *Genesis*, p. 24.

[9]Ibid.

[10]Ibid.

[11]Ibid.

[12]Ibid., pp. 22-23.

[13]As cited in Thomas Henry Louis Parker, *John Calvin: A Biography* (Louisville, KY: Westminster John Knox, 1975), p. 153.

[14]Bernard Cottret, *Calvin: A Biography* (Grand Rapids: Eerdmans, 1995), p. 225.

[15]Augustine, *The Literal Meaning of Genesis, Volume 1*, ed. John Hammond Taylor (Mahwah, NJ: Paulist Press, 1982), p. 43.

[16]Alister McGrath, *A Fine-Tuned Universe: The Quest for God in Science and Theology: The Gifford Lectures* (Louisville, KY: Westminster John Knox, 2009), p. 107.

[17]Ibid.

[18]Ira Robinson, "'Practically, I Am a Fundamentalist': Twentieth-Century Orthodox Jews Contend with Evolution and Its Implications," in *Jewish Tradition and the Challenge of Darwinism*, ed. G. N. Cantor and Marc Swetlitz (Chicago: University of Chicago Press, 2006), pp. 77-78.

[19]Judith Raskin, ed., "Maimonides," in *The Cambridge Dictionary of Judaism and Jewish Culture* (Cambridge: Cambridge University Press, 2011), p. 411.

[20]I have based this discussion of Maimonides largely on the chapters gathered in Kenneth Seeskin, ed., *The Cambridge Companion to Maimonides* (Cambridge: Cambridge University Press, 2005).

[21]Moses Maimonides, *The Guide of the Perplexed, Volume Two*, trans. Shlomo Pines (Chicago: University of Chicago Press, 1963), pp. 293-94.

[22]Moses Maimonides, *A Maimonides Reader*, ed. Isadore Twersky (Springfield, NJ: Behrman House, 1972), p. 28.

[23]Gad Freudenthal, "Maimonides' Philosophy of Science," in *The Cambridge Companion to Maimonides*, p. 158.

[24]T. M. Rudavsky, "The Impact of Scholasticism upon Jewish Philosophy in the Fourteenth and Fifteenth Centuries," in *The Cambridge Companion to Medieval Philosophy*, ed. Daniel H. Frank and Oliver Leaman (Cambridge: Cambridge University Press, 2003), pp. 346-47.

[25]Tamar Rudavsky, *Maimonides* (Sussex: Wiley-Blackwell, 2010), pp. 15-16.

[26]Maimonides, *A Maimonides Reader*, p. 28.

Chapter Nine

[1]Nancey Murphy, *Beyond Liberalism and Fundamentalism: How Modern and Postmodern Philosophy Set the Theological Agenda* (Harrisburg, PA: Continuum/Trinity Press International, 1996), p. 117.

[2]Flannery O'Connor, *The Habit of Being: Letters of Flannery O'Connor* (New York: Farrar, Straus and Giroux, 1979), p. 97.

[3]Gabriel Almond, R. Scott Appleby and Emmanuel Sivan, *Strong Religion: The Rise of Fundamentalisms Around the World* (Chicago: University of Chicago Press, 2003), pp. 82-83.

[4]A five-year study by the Barna Group released in 2011 of more than 1,200 young adults between 18 and 29 who were actively involved in their churches during their teen years but stopped attending found that one of the primary reasons many became disillusioned with church life was that they perceived their congregations as antagonistic to science. One third (35%) of respondents, for example, said that when it came to science, "Christians are too confident they know all the answers." See David Kinnaman, *You Lost Me: Why Young Christians Are Leaving the Church and Rethinking Faith* (Grand Rapids: Baker Books, 2011), p. 137.

[5]Albert Camus, *The Rebel: An Essay on Man in Revolt* (New York: Vintage Books, 1991), p. 69.

[6]Fernando Canale, "Evolution, Theology, and Method, Part 2: Scientific Method and Evolution," *Andrews University Seminary Studies* 41, no. 2 (2003): 182.

[7]N. T. Wright, *Christian Origins and the Question of God, Volume One: The New Testament and the People of God* (Minneapolis: Fortress Press, 1992), p. 35.

[8]Ibid.

[9]Ibid., p. 36.

CHAPTER TEN

[1]As cited in Charles Taliaferro and Jil Evans, *The Image in Mind: Theism, Naturalism, and the Imagination* (New York: Continuum, 2011), pp. 150-51.

[2]Richard Dawkins, *River Out of Eden: A Darwinian View* (New York: Basic Books, 1995), p. 132.

[3]John Haught, *Making Sense of Evolution: Darwin, God, and the Drama of Life* (Louisville, KY: Westminster John Knox, 2010), p. 63.

[4]C. S. Lewis, *The Screwtape Letters* (San Francisco: HarperCollins, 1942), p. 37.

[5]Frank Lorey, "Tree Rings and Biblical Chronology," *Acts and Facts* 23, no. 4 (1996), http://www.icr.org/article/tree-rings-biblical-chronology/.

[6]As cited in Karl Giberson and Donald A. Yerxa, *Species of Origins: America's Search for a Creation Story* (Lanham, MD: Rowman and Littlefield, 2002), p. 96.

[7]Friedrich Nietzsche, "Pity for Animals," in *Animal Rights: A Historical Anthology*, ed. Andrew Linzey and Paul Clarke (New York: Columbia University Press, 1990), p. 148.

[8]William Dembski, *The End of Christianity: Finding a Good God in an Evil World* (Nashville: B&H Publishing Group, 2009), p. 50.

[9]"A Statement on the Biblical Doctrine of Creation," voted by the Andrews University Seminary faculty, April 30, 2010, p. 4.

[10]Robert Hughes and J. Carl Laney, *Tyndale Concise Bible Commentary* (Wheaton, IL: Tyndale House, 2001), p. xix.

[11]Nicholas Miller, "The 'Found' World of Genesis 1," on Memory, Meaning & Faith, www.memorymeaningfaith.org/blog/2011/01/found-world-of-genesis-1-theistic-evolution-1.html.

[12]See Clyde A. Holbrook, *Jonathan Edwards: The Valley and Nature: An Interpretive Essay* (London: Bucknell University Press, 1987), pp. 73-75.

[13]For discussion of such notions from the world of "folk" religion, see my article, "True Blood: Race, Science, and Early Adventist Amalgamation Theory Revisited," *Spectrum Magazine* 38, no. 4 (Fall 2010).

Chapter Eleven

[1]C. S. Lewis, *The Problem of Pain* (San Francisco: HarperCollins, 1940), p. 137.

[2]David Foster Wallace, *Consider the Lobster and Other Essays* (New York: Little, Brown and Company, 2006), p. 251.

[3]As cited in James Rachels, *Created from Animals: The Moral Implications of Darwinism* (Oxford: Oxford University Press, 1990), p. 130.

[4]Ibid., p. 131.

[5]C. S. Lewis, *The Abolition of Man* (San Francisco: HarperCollins, 1944), p. 70; see also Andrew Linzey, *Animal Theology* (Urbana: University of Illinois Press, 1994), pp. 35-36, 153.

[6]Lewis, *The Problem of Pain*, p. 140.

[7]Ibid., p. 139.

[8]N. T. Wright, *Evil and the Justice of God* (Downers Grove, IL: InterVarsity Press, 2006), pp. 111, 113.

[9]John Stott, *Understanding the Bible* (Grand Rapids: Zondervan, 1999), pp. 54-55.

Chapter Twelve

[1]See, for example, Gregory Boyd, *Satan and the Problem of Evil: Constructing a Trinitarian Warfare Theodicy* (Downers Grove, IL: InterVarsity Press, 2001).

[2]John Polkinghorne, *Exploring Reality: The Intertwining of Science and Religion* (New Haven, CT: Yale University Press, 2005), p. 140.

[3]Wendell Berry, "Christianity and the Survival of Creation," in *Sex, Economy,*

Freedom & Community (New York: Pantheon Books, 1992), p. 97.

[4]Fyodor Dostoevsky, *The Brothers Karamazov*, trans. Richard Pevear and Larissa Volokhonsky (New York: Everyman's Library, 1992), pp. 236-46.

[5]The practice of *targum* is both explained and modeled in Brian J. Walsh and Sylvia C. Keesmaat, *Colossians Remixed: Subverting the Empire* (Downers Grove, IL: IVP Academic, 2004), pp. 38-41.

[6]William Brown, *The Seven Pillars of Creation: The Bible, Science, and the Ecology of Wonder* (Oxford: Oxford University Press, 2010), pp. 138-40.

[7]Kathryn Schifferdecker, *Out of the Whirlwind: Creation Theology in the Book of Job* (Cambridge, MA: Harvard University Press, 2008), p. 123.

[8]Robert Alter, *The Wisdom Books: Job, Proverbs, and Ecclesiastes: A Translation with Commentary* (New York: W. W. Norton, 2010), p. 177.

[9]Schifferdecker, *Out of the Whirlwind*, p. 125.

CHAPTER THIRTEEN

[1]Slavoj Žižek, "The Fear of Four Words: A Modest Plea for the Hegelian Reading of Christianity," in *The Monstrosity of Christ: Paradox or Dialectic*, ed. Creston Davis (Cambridge, MA: MIT Press, 2009), p. 53.

[2]Ibid., p. 56.

[3]See Nancey Murphy, *Theology in the Age of Scientific Reasoning* (Ithaca, NY: Cornell University Press, 1990), pp. 178-83; and Nancey Murphy and George F. R. Ellis, *On the Moral Nature of the Universe* (Minneapolis: Augsburg Fortress, 1996), pp. 184-87.

[4]John Polkinghorne, *Faith, Science, and Understanding* (New Haven, CT: Yale University Press, 2000), p. 126.

[5]Dietrich Bonhoeffer, *Creation and Fall: Two Biblical Studies* (New York: Touchstone, 1959), p. 35.

[6]John Haught, *Making Sense of Evolution: Darwin, God, and the Drama of Life* (Louisville, KY: Westminster John Knox, 2010), p. 65.

[7]Terence E. Fretheim, *Creation Untamed: The Bible, God, and Natural Disasters* (Grand Rapids: Baker Academic, 2010), p. 53.

[8]Conor Cunningham, *Darwin's Pious Idea: Why the Ultra-Darwinists and the Creationists Both Get It Wrong* (Grand Rapids: Eerdmans, 2010), p. 177.

[9]Ibid., pp. 386-87.

CHAPTER FOURTEEN

[1]See Carl Feit, "Modern Orthodoxy and Evolution: The Models of Rabbi J. B. So-

loveitchik and Rabbi A. I. Kook," in *Jewish Tradition and the Challenge of Darwinism*, ed. G. N. Cantor and Marc Swetlitz (Chicago: University of Chicago Press, 2006), pp. 208-24.

[2]Abraham Joshua Heschel, *The Sabbath: Its Meaning for Modern Man* (New York: Farrar, Straus and Giroux, 1951), p. 100.

[3]Ibid., p. 117n13.

[4]Here and in what follows I am heavily indebted to Ched Myers, "God Speed the Year of Jubilee!: The Biblical Vision of Sabbath Economics," *Sojourners Magazine* 27, no. 3 (May/June 1998): 24-28.

[5]Wendell Berry, "Christianity and the Survival of Creation," in *Sex, Economy, Freedom & Community* (New York: Pantheon Books, 1992), p. 94.

[6]Terence E. Fretheim, *Creation Untamed: The Bible, God, and Natural Disasters* (Grand Rapids: Baker Academic, 2010), p. 17.

[7]Wislawa Szymborska, "In Praise of Self-Deprecation," in *Sounds, Feelings, Thoughts: Seventy Poems*, trans. Magnus J. Krynski and Robert A. Maguire (Princeton, NJ: Princeton University Press, 1981), p. 189.

[8]Friedrich Nietzsche, *On the Genealogy of Morals* in *Basic Writings of Nietzsche*, trans. Walter Kaufman (New York: The Modern Library, 1967), p. 522.

Conclusion

[1]Flannery O'Connor, "Revelation," in *The Complete Short Stories of Flannery O'Connor* (New York: Farrar, Straus and Giroux, 1971), pp. 508-9.

Subject and Author Index

agnosticism, 143
Alter, Robert, 28, 156
animal predation, 13, 15-16, 19, 33-34, 74, 126, 128, 130-31, 134-36, 138, 140-41, 147, 151, 154, 174-75
animal suffering, 13-16, 19-20, 74, 94, 121, 126-28, 131, 135-43, 146-51, 158-62, 171, 174-75, 178
Aquinas, Thomas, 98, 107, 111
Aristotle, 106-9, 111
atonement, 128, 160
Augustine of Hippo, 104, 106
authoritarianism, 92, 115
Barr, James, 50-51, 56
Barth, Karl, 28, 79, 96-99, 148
beauty, 13, 29, 52, 103, 139
Berry, Wendell, 151, 173
Bonhoeffer, Dietrich, 28, 162
Boyle, Robert, 72
Brueggemann, Walter, 34, 51
Bryan, William Jennings, 18
Bultmann, Rudolf, 79
Calvin, John, 100-104
Camus, Albert, 116
Canale, Fernando, 116-17
capitalism, 28, 114, 175
Carson, Rachel, 174
Cartesian philosophy, 42-44, 142
Chesterton, G. K., 74, 167
Clark, Harold W., 57
Cohn, Norman, 91
concordism, 40, 70-71
consciousness, 140-42
cosmic conflict, 94, 140, 146-47, 150
cosmogony, 108
cosmology, 69, 102-4, 108, 163
creationism, 16-19, 40-46, 50, 57-59, 64-75, 78, 86-88, 94-97, 100, 107, 113-18, 126, 132, 136, 172, 177
cultural imperialism, 37
Cunningham, Conor, 94, 97, 163-65
curse, 13, 16, 34-35, 127, 131, 134-38, 151, 155, 173
Darwin, Charles, 27, 37, 39, 46, 56, 60, 64-69, 72, 95, 106, 110, 126-27, 140-43, 148, 151-52, 155, 163
Dawkins, Richard, 46, 106, 127
deceiver God dilemma, 131-34
deism, 162
Dembski, William, 136
Dennett, Daniel, 46, 106
Descartes, Rene, 42-43, 46, 75, 116, 142
Desiderius, Erasmus, 42
Dillard, Annie, 15
dinosaurs, 107, 138
divine curse dilemma, 134-39
Dostoevsky, Fyodor, 58, 152
Edwards, Jonathan, 137
Einstein, Albert, 55-56, 163
Ellul, Jacques, 58
empiricism, 118
Enlightenment, 46, 52, 75, 116, 162
epistemology, 21, 41-46, 62, 77, 86, 92, 99, 103, 116-20
Erikson, Erik, 82-83
ethics, 19, 39, 159n9, 170, 172, 175

evangelicals, 19, 37, 51, 85, 117, 137
evil, 13-14, 17, 20, 34-37, 54, 74, 88-91, 95, 127, 136, 138, 141-45, 150, 153, 161-65, 170
evolutionary biology, 7, 18, 39, 55, 59, 60, 65, 78, 106, 148, 175
evolutionary theory, 7-8, 19, 59-60, 64-65, 69, 106, 110, 116, 127, 138, 143
ex nihilo, 25, 100-101, 106-8, 136
Fabre, Henri J., 15
false consciousness, 81
falsification, 40, 44, 61-69, 77, 92, 110
Feuerbach, Ludwig, 27
fideism, 45, 58, 84, 133
flood (Noah's), 47-48, 57, 65, 138, 160, 163
Fontaine, Nicholas, 142
foundationalism, 41-45, 48, 75, 81, 86, 109, 115-21, 132, 148
free will, 14, 131, 135, 161
freedom, 27, 31-34, 58, 98, 103, 137, 144, 146, 148, 161-63, 170
Fretheim, Terence, 163, 173
fundamentalism, 19, 39, 76-82, 86, 107, 115, 134, 177
Gallup poll, 18
Gandhi, Mohandas, 44
genocide, 70n5, 126, 159n9
Geo-Science Research Institute, 69
gnosticism, 87-95, 129, 139, 144
Goodall, Jane, 174
Gould, Stephen J., 60
Grainger, D. H., 12
Hart, David Bentley, 15
Hasel, Gerhard F., 40, 46-47
Haught, John, 19, 129, 162
Hegel, Georg Wilhelm Friedrich, 14
Heidegger, Martin, 89
Heisenberg, Werner, 163
hell, 130, 137, 145, 155
hermeneutics, 39, 41, 45, 54, 73, 80-81, 86-87, 99, 103, 109-10, 117, 120, 131, 155, 172
Heschel, Abraham Joshua, 167
historical-critical methods, 39
Hodge, Charles, 76
Hoffer, Eric, 77
Holocaust, 17, 172
Hunsinger, George, 98
idealism, 14
identity foreclosure, 82-83, 86
infallibility, 54, 77, 147
Institute for Creation Research, 132
intelligent design, 14, 71, 136, 139, 162
Jonas, Hans, 89, 93
Judaism (Jewish faith), 49, 76, 88, 97, 106-11, 139, 147, 162, 169, 172
Jubilee economics, 169-73, 191
Kelvin, William Thomson, 65
kenosis, 159, 162, 165
Kierkegaard, Søren, 58
Kook, Abraham Isaac, 167
Kuhn, Thomas, 61-63, 72
Lakatos, Imre, 62-66, 72, 74
Leakey, Louis, 174
Leibniz, Gottfried, 13
Leviathan, 9, 153
Lewis, Alan, 37
Lewis, C. S., 87, 94, 129, 140, 143-49
liberalism (theological), 51, 76, 79, 98
literalism, 16-19, 39-45, 49-52, 56-59, 70-77, 86, 98-100, 109, 117, 126, 132, 134, 136, 159, 174-77
Locke, John, 28
Maimonides, Moses, 107-11, 167
Malebranche, Nicolas de, 142
Marx, Karl, 64, 81, 87, 92
materialism, 18, 38, 52, 64, 66, 92, 116, 132, 148-49, 158
Maxwell, Arthur Stanley, 17
McGrath, Alister, 38, 105-6
midrash, 34, 147

Milbank, John, 83, 85
Milton, John, 145
miracles, 74, 98-99, 101, 163
More, Thomas, 42
Morris, Henry, 18, 44, 46, 133
mortality, 16, 19, 33-35, 74, 128-30
Murphy, Nancey, 41, 113
myth, 54, 95, 98-99, 131, 144-45, 148
mythopoesis, 132
natural selection, 106, 127, 164
natural theology, 72, 161
Neurath, Otto, 120
Newton, Isaac, 55, 65-66, 72, 163
Newtonian physics, 55, 163
Nietzsche, Friedrich, 64, 89, 135, 174
nihilism, 64, 89, 116-17, 152, 155
nominalism, 133
O'Connor, Flannery, 113, 178-79
O'Donovan, Oliver, 78
Pagels, Elaine, 93
paradigms, 19, 59, 61-75, 116-17, 135, 139, 159, 163
Pascal, Blaise, 58
Pellicani, Luciano, 87, 89-90, 92
Philo of Alexandria, 108n7
Plantinga, Alvin, 51
Plato, 42, 95, 107
Polkinghorne, John, 150, 161-62
Popper, Karl, 61-63
postfoundationalism, 21, 118-21
postmodernism, 62, 116-19, 144
Price, George McCready, 18, 68, 76, 174
prophecy, 165
providence, 110, 163
quantum physics, 163
Rachels, James, 143
relativism, 62, 116-18
research programs, 61-66, 68, 72, 74-75, 139
sabbath, 37, 45, 165-75
Schifferdecker, Kathryn, 155-56
scientism, 47, 58, 64
sentience, 140-42
Servetus, Michael, 103-4
Seventh-day Adventism, 17, 54
sola scriptura, 83-85, 100, 109
stasis dilemma, 128-31
Stott, John, 37, 147
subjectivity, 42, 140
supernatural, 35, 65, 70, 94, 136, 139, 163
surrealism, 92, 129, 133
Szymborska, Wislawa, 125, 174
targum, 155
Tempier, Stephen, 111
theodicy, 17-20, 74, 94, 121, 126, 128, 131, 134, 136, 140, 143, 145, 147, 150, 155, 159, 175
theory of relativity, 55
Toulmin, Stephen, 43
ultra-Darwinism, 46, 64, 95, 126
uncertainty principle, 163
Voegelin, Eric, 89, 92
Voltaire, 13
Wallace, David Foster, 141
Walton, John, 19, 37-38
Wesley, John, 21
Wesleyan Quadrilateral, 21
Westphal, Merald, 41
Whitehead, Alfred North, 97
wildness, 32-34, 153
Wright, N. T., 84-85, 118, 144
Zimbabwe, 11-12, 166
Žižek, Slavoj, 158